V International Scientific Conference on Regional Problems of Earth Remote Sensing (RPERS 2018)

E3S Web of Conferences Volume 75 (2019)

Krasnoyarsk City, Russia
11 - 14 September 2018

Editors:

G.M. Tsibulsky
M.V. Noskov
E.A. Lupyan
L.G. Sverdlik
A.K. Sagatelian
V.I. Kharuk
V.B. Kashkin

ISBN: 978-1-5108-7907-2

Printed from e-media with permission by:

Curran Associates, Inc.
57 Morehouse Lane
Red Hook, NY 12571

Some format issues inherent in the e-media version may also appear in this print version.

No changes have been made to the content of these proceedings. There may be changes to pagination and minor adjustments for aesthetics.

Printed by Curran Associates, Inc. (2019)

For additional information, please contact EDP Sciences – Web of Conferences at the address below.

EDP Sciences – Web of Conferences
17, Avenue du Hoggar
Parc d'Activité de Courtabœuf
BP 112
F-91944 Les Ulis Cedex A
France

Phone: +33 (0) 1 69 18 75 75
Fax: +33 (0) 1 69 28 84 91

contact-edps@webofconferences.org

Additional copies of this publication are available from:

Curran Associates, Inc.
57 Morehouse Lane
Red Hook, NY 12571 USA
Phone: 845-758-0400
Fax: 845-758-2633
Email: curran@proceedings.com
Web: www.proceedings.com

TABLE OF CONTENTS

edp sciences

Web of Conferences

Statement of Peer review

In submitting conference proceedings to *Web of Conferences*, the editors of the proceedings certify to the Publisher that

1. They adhere to its **Policy on Publishing Integrity** in order to safeguard good scientific practice in publishing.
2. All articles have been subjected to peer review administered by the proceedings editors.
3. Reviews have been conducted by expert referees, who have been requested to provide unbiased and constructive comments aimed, whenever possible, at improving the work.
4. Proceedings editors have taken all reasonable steps to ensure the quality of the materials they publish and their decision to accept or reject a paper for publication has been based only on the merits of the work and the relevance to the journal.

Title, date and place of the conference

Regional Problems of Earth Remote Sensing "RPERS 2018"
September 11-14, 2018
Institute of Space and Information Technology Siberian Federal University (Krasnoyarsk City, Russia)

Proceedings editor(s):

Gennady M. Tsibulsky
Mikhail V. Noskov
Evgeny A. Lupyan
Leonid G. Sverdlik
Armen K. Sogatelian
Vyacheslav I. Kharuk
Valentin B. Kashkin

Date and editor's signature

17, avenue du Hoggar - PA de Courtabœuf – BP 112 - 91944 Les Ulis Cedex A (France)
Tél. : 33 (0)1 69 18 75 75 - Fax : 33(0)1 69 07 45 17 – www.edpsciences.org

The dynamic model of agricultural land structure on the space images in the precision agriculture tasks

Ruslan V. Brezhnev[1*] and *Yuriy* A. Maglinets[1]

[1] The Institute of Space and Information Technology of Siberian Federal University 660074, Russian Federation, Krasnoyarsk, Akademika Kirenskogo Str. 26/1

Abstract. The study is devoted to the analysis of specific features of spacial objects referred to "agricultural land" class in Central Siberia according to the results of the Earth remote sensing from the space for information support in the precision agriculture tasks. The subject of the study is temporal variability of spectral, textural and geometrical features of a land area with homogenous vegetation (hereinafter agricultural contour). During the vegetation period the agricultural contour is subject to changes caused by a combination of natural and antropogenic factors. These factors are the result of the natural course of vegetation (change of phenological phases), weather conditions and agricultural engineering measures implemented. They typically cause the change of the spacial structure of the agricultural contour resulting in non-homogenous vegetation of an agricultural crop within the agricultural contour.

1 Introduction

Aerospace monitoring methods have become widely used in agricultural activity. It applies to the whole series of thematic tasks: estimation of crop rotation field state and utilization, parameters of soil fertility and degradation, state of vegetation cover on plough lands, deposits, hay fields and grazing lands, crop yield forecasting, etc.

One of the most perspective areas in which the results of space monitoring can be used in agriculture is precision agriculture, according to which an agricultural land is treated as a special object with non-homogeneous dynamically changing structure [1, 2].

Localization of inhomogeneities and determination of a state thereof, together with the data of field studies and information on chemical composition of soils, is important for the support of farmers when making a decision on spotted application of specialized technical means for differentiated treatment of the required non-homogeneous land areas depending on their state.

[*] Corresponding author: brezhnev.ruslan@gmail.com

2 The dynamic model of agricultural land structure

A state of an F-object means a result of influence exerted by numerous regional conditions of natural and climatic and antropogenic character determining the development of an agricultural crop k^i, possessing its own temporal development model *Tm*. On a digital space image, a state of an object means values of spectral features calculated within the defined time intervals and their reference values and context-dependent descriptions that enable to correlate a localized land area w_i of the image area Θ_i with the state $f^i \in F$.

From the point of view of precision agriculture, a state of a facility cannot be considered exclusive of its local states caused by the deviation of a reference development pathway of the object defined by a time sequence of change of phenological phases – $\varphi = \{\varphi^i\}$ and agro-engineering measures (AEM) – $E = \{e^i\}$. Phenological phases define a set of states f^{vi} and AEM – f^{wi}. Thus, every state $f^i \in F$ is a composite one and can be presented in the form of a state and transition diagram G_i:

$$G_i = (F, R), \qquad (1)$$

where *F* is a set of states of the object:

$$F = f^{vi} \cup f^{wi}, \qquad (2)$$

R is a set of state transitions:

$$R = r^1 \cup r^2 \cup r^3, \qquad (3)$$

r^1 is a set of operators describing a natural change of phenological phases of agricultural crop development in time; r^2 is a set of operators describing abnormal features in the development of an agricultural crop; and r^3 is a set of operators modelling the implementation of AEM.

In the process of space monitoring organization, NDVI spectral index is used to qualitatively characterize a change in the vegetation activity and to establish bonds between index values fixed in the image and elements of the set F. Figure 1 gives an example of an agricultural land object with non-homogeneous structure shot at different time periods.

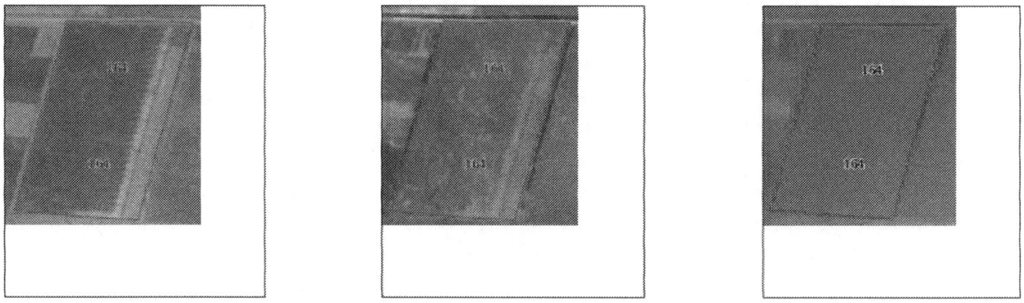

21.05.2018 25.06.2018 03.07.2018

Fig. 1. An example of an agricultural land object in the space image Sentinel-2A. Object No. 164 – oats.

Figure 2 gives a fragment of a state change and transition diagram of the object under study in time.

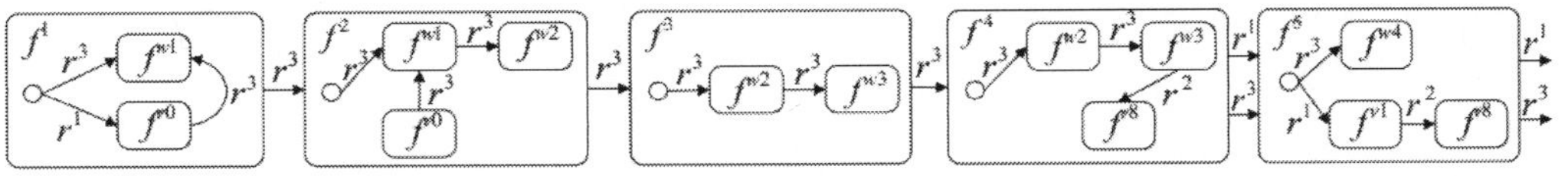

Fig. 2. A state change diagram of the object under study in time under the influence of natural and antropogenic factors.

The diagram depicts all possible states of the object corresponding to the reference time intervals. Thus, a composite state f^1, describing the object by the moment of sowing season beginning, includes the states f^{v0} and f^{w1}, meaning that in the conditions of natural development of r^1 the object will turn to the state of overgrowing, and in case of AEM implementation r^3 it will turn to the state of harrowing, accordingly, which will be definitely reflected on the values of the spectral index.

Prior to the interpretation of a state of a field corresponding to Θ_i characterized by internal inhomogeneity, it is necessary to separate Θ_i into segments w_i meeting the established criteria of homogeneity formed based on the values of NDVI:

$$s(x,y) = \begin{cases} \lambda_1 = f^{w1} \; or \; f^{w6} \; where \; 0.025 \le f(x,y) \le 0.26, \\ \lambda_2 = f^{w2} \; or \; f^{w3} \; where \; 0.26 < f(x,y) \le 0.43, \\ \dots \\ \lambda_6 = f^{v4} \; or \; f^{v5} \; where \; 0.72 < f(x,y) \le 0.85 \end{cases} \quad (4)$$

where $f(x,y)$ is a brightness function of the original image, $s(x,y)$ is a segmented image, and λ_i are segment marks, i = 0, 1, 2, …, 6.

The result of the segmentation is given in Figure 3 in the form of index images of the object under study.

21.05.2018 25.06.2018 03.07.2018

Fig. 3. Index image of the object No. 164.

As distinct from Θ_i spatial coordinates of which are fixed, spatial position of w_i shall be localized. For this purpose segmented areas w_i are transformed from an iconic level to a level of geometrical and relational structures, where a feature vector formed for w_i includes not only iconic features, but also metric characteristics physical interpretation of which enables us to describe a position, dimensions and form of inhomogeneities within the field (Fig. 4). This way w_i can be considered and analyzed as an independent object.

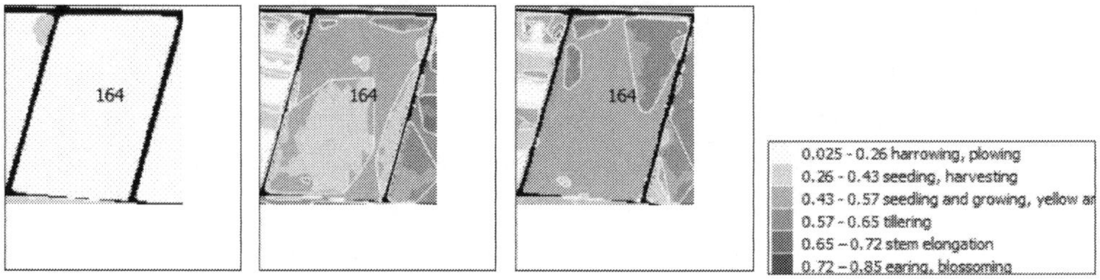

21.05.2018 25.06.2018 03.07.2018

Fig. 4. Segmented image of the object No. 164.

Metric characteristics are as follows:

1. Geographical coordinates in the given projection $P_{\Pi} = \{(x_i, y_i)\}$, describing the location of w_i.
2. Area of the object with polygonal geometric pattern:

$$NS = \frac{1}{2}\left|\sum_{i=1}^{n-1} x_i y_{i+1} + x_n y_1 - \sum_{i=1}^{n-1} x_{i+1} y_i - x_1 y_n\right|, \quad (5)$$

where (x_i, y_i), $i = 1, 2, \ldots, n$ is a sequence of coordinates of the neighboring vertices of the polygon, $(x_{i+1}, y_{i+1}) = (x_1, y_1)$.

3. Perimeter of the object:

$$NP = \sum_{i=1}^{n} \sqrt{(x_{i+1} - x_i)^2 + (y_{i+1} - y_i)^2}, \quad (6)$$

where $i = 1, 2, \ldots, n$ is a number of vertices of the contour.

4. Thickness of the object:

$$T = 4\pi(NS/NP^2), \quad (7)$$

Thickness is a quantitative index characterizing the form of a non-heterogeneity contour. Under the ideal conditions, the form of the contour w_i shall be convex, i.e. the value of T shall be close to 1. Yet, in practice heterogeneities possess borders with stretched broken form, i.e. T tends to 0.

In a result, an accompanying task emerges - to bring the inhomogeneity contour to the convex form, which in practice enables to organize parallel movement of agricultural machinery equipped with spotted treatment devices.

Bringing the contour to the convex form is based on Jarvis's Algorithm, which lies in the determination of a sequence of elements of a set of contour points forming a convex envelope for this set. The result of such conversion is given in Figure 4 during the analysis of object inhomogeneities for 25.06.2018 and 03.07.2018. The conversion covers only those non-homogeneous areas which, from the farmer's point of view, are appropriate for spotted treatment [3]. For this purpose, it is firstly necessary to take into account the minimum limitation of the non-heterogeneity area – NS^*_{wi} which shall include the requirements for the width of suspended equipment used. Secondly, the accuracy of determining the belonging of a non-homogeneous area to this or that state is very important in the process of information support rendering.

To do that, it is firstly necessary to assess the deviation of index values of the non-homogeneous area w_i from the defined norm. Mean square deviation σ is calculated for each inhomogeneity. After that, the probability of correct state determination is calculated. This calculation is based on the probability of mean index value $\bar{N}$ within the area w_i falling into the diapason of values corresponding to the reference states of the object (4) typical of the current time *Tm* and is calculated using a formula of probability of a discrete random variable falling into the interval:

$$P(\alpha < \bar{N} \leq \beta) = \Phi_o\left(\frac{\beta - d}{\sigma}\right) - \Phi_o\left(\frac{\alpha - d}{\sigma}\right), \tag{8}$$

where α and β are the defined threshold values of the intervals, d is mathematical expectation, Φ_o is Laplace's function, $\bar{N}$ is calculated as follows:

$$\bar{N}(w_i) = \frac{1}{n}\sum_{i=1}^{n} NDVI(x_i, y_i), \tag{9}$$

where $NDVI(x_i, y_i)$ is a brightness function of the index image point.

In case σ value for the non-homogeneous area w_i is insignificant, the probability of this area falling into the diapason corresponding to a particular state is high and close to 1. Therefore, the exactness of current actual state determination f_{Aj}^i ($f_{Aj}^i \in F$) is close to 100%. After that, current state of the area w_i can be compared to a reference state. In case current state does not coincide with the reference one, whereby it is highly probable that the state of w_i is determined correctly, it is recommended to examine the area w_i under study.

In case σ value for the non-homogeneous area w_i is significant, the correctness of such area determination reduces greatly and an error increases.

In a result of the described analysis of a non-homogeneous dynamically changing structure of an agricultural land object throughout the whole vegetation period based on the space images, object-relational structures of an inhomogeneity presented in the form of inhomogeneity maps are formed (Fig. 4). Whereby, each non-homogeneous area w_i of the j-th object is described using the following set of features enabling to compare a physical state of thef^i field in the localized area w_i in the image:

$$w_i = < ID, \{P_{\Pi i}^i\}, NS_j^i, NP_j^i, T_j^i, \bar{N}_j^i, Tm, k_j^i, e_j^i, \varphi_j^i, f_{Aj}^i >, \tag{10}$$

where *ID* is a number or name of the agricultural land object.

3 Conclusion

A graph model of a spatial object "agricultural land" reflecting a variety of states and transitions caused by the abovementioned factors has been designed. The model accounts for the planned and actual development path of the object in time. It contains means of expression that describe the inhomogeneity of the object in time and space. The model has been used as the basis for agricultural monitoring system developed by the authors [4, 5]. This system allows monitoring of the seasonal vegetation, localizing abnormal features in plants development, and monitoring of agro-engineering measures and their results to ensure information support of end users (agronomists, heads of agricultural enterprises and authorities). Based on this model, the system generates maps of inhomogeneities for each field at the set monitoring rate which allows for a comprehensive assessment of the object in time and timely revealing of adverse developments.

The reported study was funded by Russian Foundation for Basic Research (No.18-47-242002), Government of Krasnoyarsk Territory, Krasnoyarsk Regional Fund of Science, to the research project: «The development of technology for creating intelligent information systems of object-oriented monitoring of territories based on remote sensing data».

https://doi.org/10.1051/e3sconf/20197501001

References

1. R. V. Brezhnev, Yu. A. Maglinets, K. V. Raevich, G. M. Tsybulsky, II International Conference: Information technology and nanotechnology, **Section 2**, 1090 (2018)
2. R. V. Brezhnev, Yu. A. Maglinets, CEUR Workshop Proceedings, **2033**, 247 (2017)
3. R. V. Brezhnev, Yu. A. Maglinets, IV International Scientific Conference: Regional Problem of Earth Remote Sensing, 36 (2017)
4. R. V. Brezhnev, E. A. Mal'tsev, Pattern Recognition and Image Analysis, **25**, 201 (2015)
5. R. V. Brezhnev, Yu. A. Maglinets, Journal of Sib. Fed. Univ. Engin. & Techn., **10**, 819 (2017)

Research of efficiency of the statistical non-parametric pattern recognition models for forest land classification

Aleksander P. Guk [1,*] , *Larisa G.* Evstratova [2]

[1] Siberian State University of Geosystems and Technologies, 630018, 10 Plakhotnogo Str., Novosibirsk, Russian Federation
[2] State University of Land Use Planning, 105064, 15 Kazakova str, Moscow, Russian Federation

Abstract. The principles of the creation of pattern recognition models, based on using multispectral imagery of forest land, have been analyzed. The statistical non-parametric model has been suggested as a basic pattern recognition model and the probability density function – as a recognition feature. Efficiency of the different quality criteria has been discussed. The main directions for improving the pattern recognition models are regarded.

Measured spectral brightness of objects recorded as brightness of image elements is key information for pattern recognition using multispectral images. As a result of processing these measurements, it is required to classify the objects and determine their qualitative characteristics. To directly determine these characteristics is not possible, therefore, based on the measurements, it is necessary to form features that uniquely determine the object and some of its properties. Thus, the object and its state are associated with measurements through the selected characteristics. It would be ideal to take measurements directly. However, the state-of-the-art Earth remote-sensing instruments do not allow doing it. Moreover, it is difficult to measure even the spectral characteristics of an object, since measurements are influenced by a lot of factors (atmosphere, location of the carrier and the survey system). Another important factor is that spectral reflectance of various objects is strongly correlated and fluctuated.

Thus, a complex problem arises that includes several stages: study of spectral properties of object reflection, description of object properties based on spectral brightness coefficients (formation of object features from spectral brightness coefficients), and development of a communication pattern between the measured brightness and features. As is obvious, to implement this scheme it is needed to study the object properties, to assess the influence of the atmosphere and other factors and to form models describing all these relations. Obviously these are probabilistic-deterministic models (deterministic chaotic models), the solution of which presents severe difficulties.

However, in practice, a different approach is used. A model is created that connects measurements and properties and it is assumed that the properties are selected in such a

* Corresponding author: guk_ssga@mail.ru

way that they uniquely determine an object. Often these are very simple models for interactive decryption that use the intuition of the decrypting operator. They are cluster analysis models. Statistical models such as the Bayesian classifier, the maximum likelihood method, etc. are also among the same type of models. To simplify the problem solution the majority of such models use the normal law. However this often leads to an inadequate result.

In [1, 2] a statistical nonparametric approach to deciphering forest tracts based on multispectral satellite imagery is suggested. The essence of the suggested approach is as follows. For all classes of objects that need to be recognized by their images on multispectral imagery reference features are created – the probability density function for the brightness of the corresponding elements of a multispectral image. Reference probability density functions are defined by sufficiently large samples of size $N \leq N_{min}$. Measurements are performed from the images of the reference objects; the number of measurements N_{min} is taken with the certainty matching the criterion $\alpha \leq 0.1$.

As is known, with non-parametric methods, the probability density functions are defined as a result of statistical studies by various methods (histogram, Parzen method, polygonal estimates, etc.).

Taking into account that, when processing images, a random function is given by discrete values with a given step on the bounded interval [0 - 255], [0 - 1023] or [0 - 4095], then the histogram method should be considered as the most acceptable for estimating the non-parametric probability density function. Indeed, all other methods are based on the approximation (interpolation) of sample values at some interval, which is useful for a large number of measurements of the function being evaluated.

The histogram approach allows us to obtain a discrete estimate of the probability density functions, the certainty of which depends on the sample size. In [2], as a result of the research, the values of the minimum sample size that can be used to classify the data are obtained (with the certainty 1- α, where $\alpha = 0.05$).

When assessing the reference probability density functions, it is necessary to check the distribution "for normality". If the distribution does not conform to the normal law, then it is advisable to use non-parametric statistics. Normality test is based on the criterion ω^2 [3]:

$$\omega_m^2 = m\int....\int \psi^2\left[F(X_m,\mu)\right]\left[F_m(X_m) - F_m(X_m,\mu)\right]^2 dF(X_m,\mu) > \omega_\alpha^2 \qquad (1)$$

where $F_m(X_m)$ is the empirical probability density function of the sample $\{x_i\}^m$; $F_m(X_m,\mu)$ the normal probability density function with the parametric variable μ and $\Psi^2[F]$ is the weight function.

The reference functions are different for multispectral images obtained by different survey systems. Thus, for using the non-parametric approach it is required to create a data bank that includes probability density functions for all classes of objects that will be deciphered in the images. These functions must be obtained for different types of survey systems as well. As for the scale of the images, as a result of the research [2], we found out that when the image scale is changed within ±30%, the probability density function changes within the accuracy of the method.

After creating a bank of reference functions, you should decipher the images. Image segmentation is done for this. It is advisable to use area attributes, for example, textural ones. The number of elements in the selected segments must be at least n_0 and is determined from the condition of the minimum possible size of the section for which the difference from the reference distributions does not exceed $\alpha \leq 0.05$. Then, for each selected segment, the histogram method is used to obtain estimates of the probability density function, which are compared with all the reference functions.

Thus, the essence of the method described is to obtain reference probability density functions, to compute the match value of the probability density function obtained by measuring the object image in the deciphered image and assigning the object to a particular class.

The key problem of this method is to choose the decision rule and the quality criterion of the object function.

As is known [4], all types of decision rules are based on the formation of the likelihood ratio and its comparison with the certain threshold *c*, the value of which is selected in accordance with the quality criterion:

$$L = \frac{F(x_1, x_2 ... x_n / s_1)}{F(x_1, x_2 ... x_n / s_0)} \geq c \tag{2}$$

Practically when calculating statistics, it is necessary to take into account the form and structure of the probability density function.

Accordingly, statistics are divided into [4]:

a) Kolmogorov statistics, which estimate the maximum discrepancy between the values of the measured and reference functions:

$$D = \sup_x \; [F_0(x) - F_j(x)] \tag{3}$$

b) weighted Kolmogorov statistics of the form:

$$D = \int_{-\infty}^{+\infty} [F_0(x) - F_j(x)]^2 dF_0(x) \tag{4}$$

when more weight is given to the distribution tail area;

c) statistics of the form χ^2 (chi square):

$$D = \sum_{x_k}^{x_k + \Delta_l} \frac{F(x) - F_0(x)}{F(x)} \tag{5}$$

where *k* and *l* are quantities that allocate the necessary intervals from the probability density functions for quality analysis of the functions; $F(x_0)$ and $F(x)$ are the reference and measured probability density functions, respectively.

These statistics are easy to use for a univariate probability density function and the computational problem becomes quite complicated with increasing dimensionality of the measurement vector and the number of recognized objects. Moreover it is difficult to interpret the result obtained.

The probability density function has the properties of both element-by-element classification and area classification. It all depends on the choice of the parameters of the quality criterion. Indeed, if in (4) we properly choose the weights and integration limits, and the parameters *k* and *l* in the estimate (5), then we can select one or another part of the brightness as the main feature in recognition.

In the first stage, when using the suggested method [1], instead of statistics of the type "a", statistics based on definition of the correlation function $R_j(0) = \frac{F_0(x) F_j(x)}{F_0(x)}$ and the definition of $\max Rj_{\forall j}$ were used. This criterion, at least, allows us, in some sense, to estimate the general discrepancy of the probability density functions.

Approximately the same result is determined by the Fisher correlation method *Z* [3].

The recognition method suggested in [2] used the measurements of only one channel, which are completely characterized by a univariate probability density function. The channel informativity was determined visually – by the maximum difference of the probability density function curves for different classes in different channels. Although this approach is not mathematically justified, in practice its application is quite acceptable [4]. Nevertheless, it is clear that in general it is necessary to use the information from all the channels of the survey system and, accordingly, it is required to estimate the multivariate density of information and to use the multivariate matching criteria.

The method of using one channel [2] to recognize the survey forest characteristics made it possible to obtain a good result (with recognition accuracy of above 90%). And although this result is obtained for certain conditions and a limited range of characteristics of objects (trees), it is possible to get a better result.

Improving stability of the algorithm and accuracy of the estimates obtained is possible if the given data structure is improved, for example, by orthogonal transformation of the reference measurement matrix and using the above criteria for multivariate probability density functions. The improved statistical model can be applied to a wide class of natural objects. In particular, the technique can be used to obtain statistical reference and for other objects, for example, agricultural crops.

The research carried out in this area has shown that the PCA transformation alters the form of the probability density function. The distribution of the first image component for each vegetation type is close to normal. To describe the first component it is entirely justified to use normal distribution and apply the corresponding simplified recognition criteria. At the same time, the third component has the greatest difference between the probability density functions of different forest types. This leads to the conclusion that it is necessary to use multivariate probability density functions or construct a decision tree classifier that uses separate image components. Note that although multivariate functions are preferable, their practical production for reference functions is associated with technical difficulties.

The empirical research has also confirmed the conclusion that spectral measurements cannot uniquely characterize the type and other characteristics of forest. By measuring the spectral characteristics only (even in the case of ideal measurements free from the influence of external factors), the type of trees is not uniquely determined by these measurements. Accordingly, the linear transformation of these spectral measurements does not save from ambiguous solution, although at the same time it identifies the most important factors affecting the image of vegetation. Thus, in future it is necessary to determine additional characteristics that would help eliminate ambiguity of object description and improve recognition accuracy.

The suggested method can be used to solve other problems. For example, we studied various spectral descriptions of the image of individual trees (using the original images or transformed by the Gaussian operator (Fig.1)). Reference images of trees help determine most of survey forest characteristics from multispectral images.

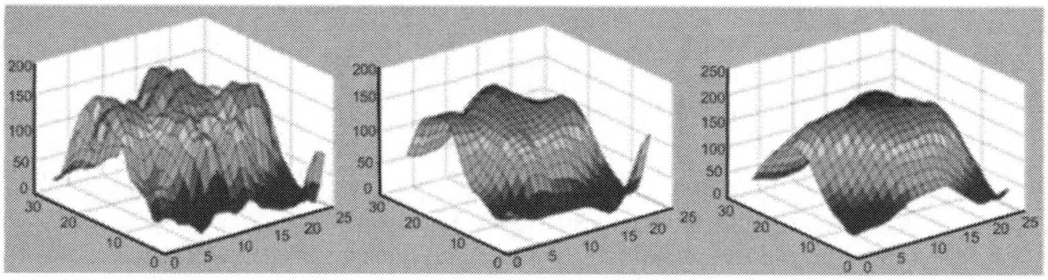

a b c

Fig. 1. Spectral tree image using the Gaussian operator with various σ:
a) the original image, b) $\sigma = 1.5$, c) $\sigma = 2.0$.

It should be noted that when using the non-parametric approach a significant part of the recognition problem solution is transferred to the segmentation process, which is based on using textural features of object images. (Note that the actual degree of homogeneity can be estimated using the probability density function.). If segmentation results in a high degree of homogeneity of the selected segment, the application of the non-parametric statistical method gives a high degree of object recognition (about 95%).

To improve the suggested approach and non-parametric statistical models it is necessary to:

1. determine the stable statistical characteristics of the brightness distribution of images of given classes of objects transformed in accordance with the a priori given probabilistic model of multispectral measurements – the method of principal components (PCA), independent components (ICA), Tasseled Cap (TC) or using other models;
2. choose the characteristic sections of the probability density function curve that determines the greatest class difference and develop appropriate criteria for model conformance evaluation;
3. choose the most informational characteristics and create effective space of features that provide the maximum for the criteria used that determine the differences in the probability density functions for the classes studied.

References

1. A.P. Guk, L.G. Evstratova, New statistical approach of forest image recognition, 14-16 (2016)
2. A.P. Guk, Izvestia vuzov Geodesy and Aerophotosurveying, Vol. **5**, 166–170 (2015)
3. K. Fukunaga, Introduction to statistical pattern recognition. (Academic Press, New York and London, 1972)
4. D.R. Cox, D.V. Hinkley, *Theoretical statistics* (Imperial college, London, 1974)

Remote Sensing Methods for the Retrieval of Inventory and Bioproductivity Parameters of Forests Using High Resolution Satellite Images

Egor Dmitriev[1,*], *Vladimir* Kozoderov[2] , *Sergey* Donskoy[3] , *Petr* Melnik[4] , and *Anton* Sokolov[5]

[1]Institute of Numerical Mathematics of Russian Academy of Sciences, ul. Gubkina 8, 119333 Moscow, Russia
[2]M.V. Lomonosov Moscow State University, Leninskiye Gory 1, 119991 Moscow, Russia
[3]Federal Forestry Agency ROSLESINFORG, Volgogradsky Prospect, 45- 1, 109316 Moscow, Russia
[4]Mytischi Branch of Bauman Moscow State Technical University, 1st Institutskaya street 1, 141005 Mytischi, Moscow region, Russia
[5]Laboratoire de PhysicoChimie de l'Atmosphère Université du Littoral Cote d'Opale, Avenue Maurice Schumann 189A, 59140 Dunkerque, France

Abstract. A method for automated processing high spatial resolution satellite images is proposed to retrieve inventory and bioproductivity parameters of forest stands. The method includes effective learning classifiers, inverse modeling, and regression modeling of the estimated parameters. Spectral and texture features are used to classify forest species. The results of test experiments for the selected area of Savvatievskoe forestry (Russia, Tver region) are presented. Accuracy estimates obtained using ground-based measurements demonstrate the effectiveness of using the proposed techniques to automate the process of updating information for the State Forest Inventory program of Russia.

1 Introduction

The State Forest Inventory (SFI) is the key element of forest inventory work. Using SFI materials in medium- and long-term planning document development will help make informed decisions on forest management. A network of permanent inventory plots of constant radius is created to determine quantitative and qualitative characteristics of forests and systematic monitoring of their health.

The materials of the forest inventory, updated for forest growth course and according to Earth remote sensing data, are used to build a network of permanent inventory plots. Last forest inventory maps being vectorized with current changes, digital base material is created for all the targets. After making the current changes from the forest inventory maps, the digital base materials are updated according to Earth remote sensing data.

In most cases when updating it is only a matter of contour satellite image interpretation and detection of unaccounted logged areas, areas of forest destruction and crowing forest fires. However, for the intensive forest management areas it is necessary to retrieve the

* Corresponding author: yegor@mail.ru

extended information on young forests, selective logging areas and forests that do not get forest inventory by using satellite images of higher resolution (up to 0.5 m).

The results presented in this paper are based on processing images of the commercial satellite system WorldView-2. The system contains an eight-channel multispectral camera with 1.84 m resolution and a panchromatic camera with 0.46 m resolution. Scientific studies presented in recent publications demonstrate that WorldView-2 images can be used effectively to determine various characteristics of stands, in particular, their leaf area index [1] and species composition [2].

Adoptions of the new Forest Code of the Russian Federation and influence of political and economic factors have dramatized the situation with relevance of forest inventory. Annual volume of forest inventory work has fallen by 80%. The issue of strategic forest management planning, as well as the sectoral reporting in the framework of international agreement implementation, including carbon sequestration, has risen sharply. Remote inventory methods make it possible to significantly increase the SFI efficiency, since currently large forest areas of the Russian Federation are not wired with an inventory plot network. Moreover, ground-based methods for forest inventory are considered time-consuming and expensive.

2 The method of remote inventory automation

Species composition and age classes of mixed stands are determined by learning classification methods. The classifier proposed in this work is an effective modification of the well-known Error Correcting Output Codes (ECOC) classifier [3]. The ECOC allows solving multi-class classification problem by a series of binary classifications and sequent voting. The general ECOC scheme is as follows. Each class is assigned a unique code (vector string) containing numbers 1, -1, and 0. Class codes have the same specified length and can be combined into a coding matrix whose strings correspond to recognizable classes, and columns – to different methods of some binary classifier's learning. To ensure that a classification problem can be successfully solved any of two classes must have a coding matrix column which these classes are separated in.

At the decoding stage binary classification problems determined by coding matrix columns are solved. Binary classifiers' responses are written into a row vector. Next the distances between the received string and the class codes in the sense of some given measure are calculated. The class corresponding to the minimum distance is the ECOC response.

Standard classification algorithms of different complexity and computational efficiency are used as basic binary classifiers. Nonlinear classifiers usually have an advantage in accuracy due to application of more complicated discriminant surfaces. At the same time, they lose in speed and sustainability of learning outcomes. The ECOC allows us to construct an effective combination of these algorithms [4].

Spectral brightness of separate image pixels normalized to integrated brightness is considered as a spectral feature. This transformation can significantly reduce the spread resulting from brightness difference. Textural features are based on the gray-level co-occurrence matrix statistics (GLCM). The GLCM is calculated for some small rectangular area of the original image centered on the classified pixel. The GLCM elements are frequencies of occurrence of brightness gradients along a given direction. Contrast, correlation, energy and uniformity are used as statistics. These statistics can be calculated for several directions.

Combination of textural and spectral features, even in the case of multispectral images, forms high dimensional space. When using standard classification methods, we inevitably come to curse of dimensionality problem or useful information loss. In this paper we propose a method that should solve this problem. The concept of this method is represented in Fig. 1.

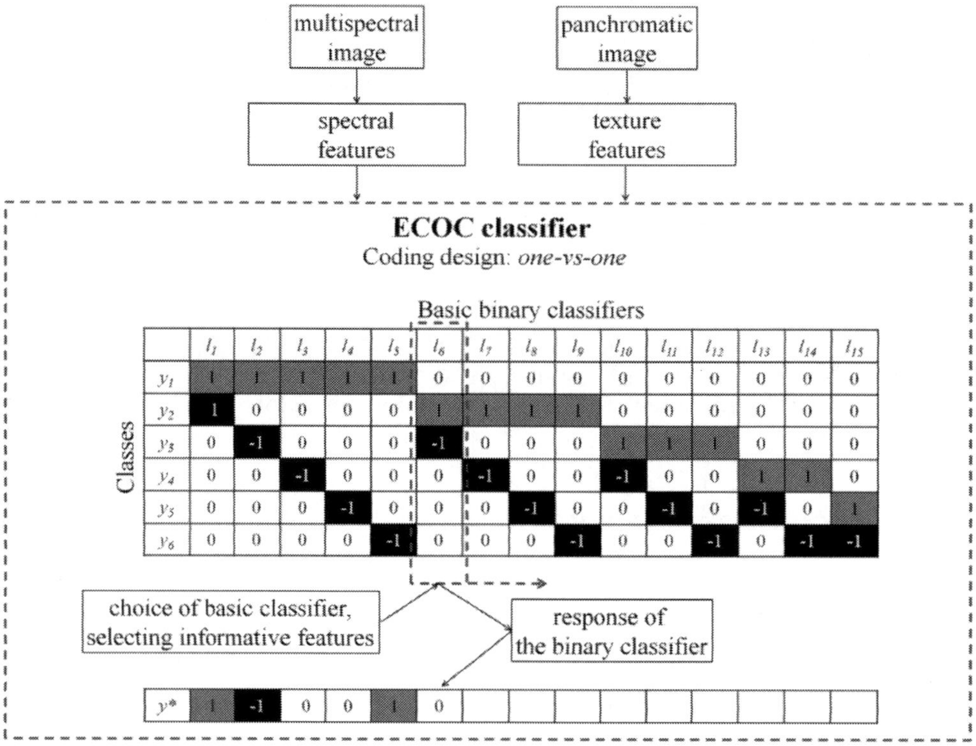

Fig. 1. The scheme of the classification method used for the recognition of ground objects based on spectral and textural features. Here y_i – names of classes, l_i – labels of binary classifiers, y^* – label of a classified object.

The ECOC uses a coding matrix corresponding to one-vs-one strategy. The rationale for this choice is presented in [4]. For each binary classification problem a basic algorithm is selected to ensure maximum classification rate without significant loss of accuracy. Basic classification algorithms should not produce significant extrapolation beyond a learning set, i.e. refused classification is realized for each of them. For each binary classification problem an optimal set of features is selected using the regularized stepwise method [5].

For the retrieval of stands' bioproductivity parameters, we use an approach based on inverse modeling of remotely measured spectral intensity of solar radiation reflected by forest canopy. Currently there are different models of forest canopy reflectivity that differ in complexity. Earlier in [6] we proposed a model of forest canopy based on its projection parameters of canopy and crown density. The model considers reflection of direct and scattered radiation from the surface of tree leaves visible from above, reflection from intercrown vegetation and multiple reflection between tree leaves and intercrown vegetation with account for shadows in the intercrown area and tree crowns.

Inverse modeling is based on the look-up table method, which contains the spectral characteristics of forest canopy with typical values of stand parameters. The projection characteristics obtained are associated with stand fraction biomass and net primary production (NPP), based on the statistical models calibrated by ground-based measurements. Particularities of the use of this model for the processing of high spatial resolution images are presented in [7].

3 Numerical experiments and discussion

The subject of research is the territory of the Savvatyevskoe forestry of the Tver region. It is located in the south-eastern part of the Tver region in the Kalinin administrative district over 15.7 thousand hectare area. The forest stands of the Savvatyevskoe forestry are referred to the mixed coniferous broadleaf forest zone, to the area of mixed coniferous broadleaf forests of the Russian Federation. The main dominant species in the forestry are Pine, Spruce and Birch.

In 2007 the Tver forest inventory enterprise (currently the Tver branch of FSBI "Roslesinforg") carried out forest inventory in this forestry. Forest inventory was conducted first-class by ocular estimate with a 1x1 km net of rides. The survey engineers did full-scale work describing stand composition, forest density, mean diameters, height, average annual stand depletion and other forestry indices. As a result of the forest inventory we have a forestry database which contains all the information about the forest health.

In 2009 SFI work was carried out in the Tver forestry of the Tver region. A series of work was done to create a digital base material, to make changes from forest inventory maps and to update the influence of natural and antropogenic factors by Earth remote sensing. Permanent inventory plots were formed and their data was compiled with the rest data from the permanent inventory plots of the Tver region and was submitted for processing.

The results of the WorldView-2 image classification of the Savvatyevskoe forestry are shown in Fig. 2. The black lines indicate the boundaries of the forest plots. The results obtained correlate well with the ground data of species composition and the visual test data of the test area objects.

The classification error estimates are presented in Table 1. The table rows correspond to true classes and the columns, with the exception of the last two, correspond to the classifier's responses. The last two columns contain relative errors obtained by the reclassification method (RCM) and by the cross-validation method (CVM). RCM is considered as a prior estimate of the classification error, since learning and test ensembles coincide in its calculation. When calculating CVM, test data is not involved in learning, and it is regarded as an independent estimate of the classification error. Close agreement of RCM and CVM indicates that there are no problems with relearning and curse of dimensionality.

It is easier to classify water bodies, soils and artificial objects. Classification error probability is less than 1%. Individual pixels of these objects can be misclassified as a pine stand. This is apparently due to the presence of spectral characteristics corresponding to significant intercrown gaps typical for mature and overmature stands. Error probability in recognition of pine stands and meadow vegetation also does not exceed 1%. Meadow vegetation can be misclassified as a pine or birch stand. It is possible to see high discriminability of coniferous stands, small-leaved stands and meadow vegetation. The greatest errors occur when recognizing aspen stands. Total misclassification probability is less than 5%.

When determining bioproductivity parameters of stands it is necessary to have information about projective characteristics and species composition as well as about age and the cite index. Age classes can also be recreated using remote sensing data; however the errors will be much higher than in tree species definition. For the remote retrieval of the cite index it is necessary to make aerial lidar measurements of crown height and diameter, otherwise the cite index would be considered unchanged and the age classes would be updated by ground data.

 https://doi.org/10.1051/e3sconf/20197501003

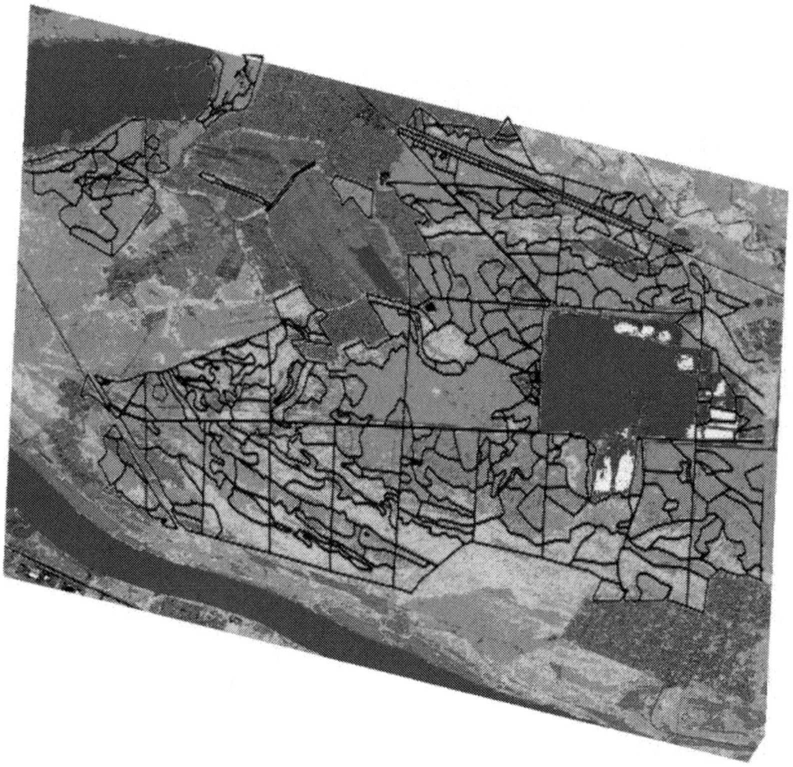

Fig. 2. The results of subject image processing WorldView-2 for the test tract in the Savvatyevskoe forestry.

Table 1. The results of classification error estimate of the WordView-2 multispectral images for basic types of objects in the test area.

	Water	Artificial	Soils	Pine	Birch	Aspen	Herbs	RCM	CVM
Water	2483	0	0	1	0	0	0	0	0.0004
Artificial	0	594	0	5	0	0	0	0	0.0083
Soils	0	0	746	2	0	0	0	0	0.0027
Pine	0	0	0	6052	43	0	11	0.0039	0.0088
Birch	0	0	0	11	1844	133	12	0.0664	0.0780
Aspen	0	0	0	0	187	795	0	0.1284	0.1904
Herbs	0	0	0	10	6	0	1888	0.0031	0.0084

The estimates of biomass give forest fractions (leaves/needles, branches, etc.)obtained for the Savvatyevskoe forestry test areas vary widely. Birch stand leaves have a 2.5-4 t/ha biomass, and pine stand needles have a 7-8 t/ha biomass. Typical averaged NPP values are 350-450 g/(m^2 year). The highest values correspond to deciduous forest stand. Forest plots with a majority of birches and high values of canopy density have NPP values of 550-600 g/(m^2 year). Pine forests have NPP from 50 to 350 g/(m^2 year). Dense stands have the highest values. Quantitative validation of biomass estimates and NPP productivity of stands is difficult. These results generally correspond to the estimates obtained by different authors at a global scale. A nuanced analysis of the bioproductivity parameter recovery results is given in [7].

4 Conclusion

Based on the analysis of the results obtained we can conclude that the proposed method for the retrieval of inventory and bioproductivity parameters of forests can be used to update attributive information on the forest stands structure taken from the permanent inventory plots. Considering the location of the Savvatyevskoe forestry, the method can be applied to other coniferous broadleaf stands during forest inventory activities within the framework of SFI in the Russian Federation.

This work was supported by the Russian Science Foundation (project No. 16-11-00007) and the RFBR (projects No. 14-05-00598 and 14-07-00141).

References

1. J. Tian, L. Wang, X. Li, H. Gong, C. Shi, R. Zhong, X. Liu, International Journal of Applied Earth Observation and Geoinformation. **61,** 22-31 (2017).
2. M. Karlson, M. Ostwald, H. Reese, H.R. Bazié, B. Tankoano, International Journal of Applied Earth Observation and Geoinformation, **50**, 80–88, (2016).
3. T.G. Dietterich, G. Bakiri, Journal of Articial Intelligence Research, **2**, 263-286 (1995).
4. .V. Dmitriev, V.V. Kozoderov, . . Dementiev, .N. Safonova, Optoelectronics, Instrumentation and Data Processing, **54(3)**, 213-221, (2018).
5. .V. Dmitriev, Earth Exploration from Space, **3**, 22-32 (2013).
6. V.V. Kozoderov, T.V. Kondranin, V.S. Kosolapov, V.A. Golovko, .V. Dmitriev, Issledovanie Zemli iz Kosmosa, **1**, 57-65 (2007) (in Russian).
7. V.V. Kozoderov, .V. Dmitriev, Issledovanie Zemli iz Kosmosa, **6**, 75-88, (2017) (in Russian).
8. A.Z. Shvidenko, S. Nilsson, V.S. Stolbovoy, et al, Russian Journal of Ecology, **2,** 83-90 (2001).
9. S.T. Gower, O. Krankina, M. Olson, M. Apps, S. Linder, C. Wang, Ecological Applications, **11,** 1395-1411 (2001).

Numerical Analysis for Problems of Remote Sensing with Random Input Data

Boris Dobronets[1,*] and *Olga* Popova[1,**]

[1]Institute of Space and Information Technology, Siberian Federal University, Kirenskogo 26, Krasnoyarsk, 660074 Russia

Abstract. The study is devoted to the remote sensing data processing using the models with random input data. In this article we propose a new approach to calculation of functions with random arguments, which is a technique of fast computations, based on the idea of parallel computations and the application of numerical probability analysis. To calculate a function with random arguments we apply one of the basic concepts of numerical probabilistic analysis as the probabilistic extension. To implement the technique of fast computations, a new method based on parallel recursive calculations is proposed.

1 Introduction

Remote sensing data analysis often includes numerical study of models with random input data. Such models are traditionally studied by Monte Carlo methods. Monte Carlo method is a powerful approach, but it has some serious shortcomings. First of all, this is an extremely low rate of convergence. Non-Monte Carlo methods have been developed since the 1960s. A major non-Monte Carlo approach is interval analysis. However, interval analysis is computed only on the boundaries of random processes without examining their internal distributions. In our work, we develop a technique that uses Numerical Probabilistic Analysis (NPA) to solve various problems with stochastic data uncertainty.

The basis of numerical probabilistic analysis is numerical operations on probability density functions of the random values. These are operations "+", "−", "·", "/", "↑", "max", "min", as well as binary relations "≤", "≥" and some others. The numerical operations of the piecewise polynomial function arithmetic constitute the major component of NPA.

Using the arithmetic of probability density functions and probabilistic extensions, we can construct numerical methods that enable us solving systems of linear and nonlinear algebraic equations with random parameter [6].

In our approach we use the piecewise polynomial models to represent probability density functions:

- piecewise constant functions (histograms);
- piecewise linear functions (frequency polygons);
- piecewise polynomial functions (splines).

*e-mail: BDobronets@yandex.ru
**e-mail: OlgaArc@yandex.ru

In many cases problems of distance sensing can be reduced to the numerical analysis of functions

$$z = f(x_1, x_2, \dots, x_n).$$

When the probability density of $x_1, x_2, \dots, x_n$ is known, Monte Carlo method is used. Using NPA for these problems is a thousand times more effective than Monte Carlo method.

2 Probabilistic extensions

One of the most important problems that NPA deals with is to construct probability density functions of random variables.

Let us start with the general case when $(x_1, \dots, x_n)$ is a system of continuous random variables with joint probability density function $p(x_1, \dots, x_n)$ and the random variable z is a function $f(x_1, \dots, x_n)$

$$z = f(x_1, \dots, x_n).$$

By *probabilistic extension* of the function f we mean a probability density function of the random variable z.

Definition. The support of $\boldsymbol{f}$, written supp($\boldsymbol{f}$), is the set of points $x \in R$ where $\boldsymbol{f}$ is non-zero

$$\text{supp}(\boldsymbol{f}) = \{x \in R | \boldsymbol{f}(x) > 0\}.$$

Consider the procedure for computing the probabilistic extensions for the one-dimensional case. Let there be given a functional dependence

$$z = f(x),$$

where x is a random variable. Let $\boldsymbol{x}$ be the probability density function of a random variable x with support $[\underline{x}, \overline{x}]$. Further $\{x_i(z) \in [\underline{x}, \overline{x}] | i = 1, \dots, n\}$ are the roots of the equation $z = f(x)$.

It is necessary to find the probability density function $\boldsymbol{f}$ of the random variable z as probabilistic extension $\boldsymbol{f}(\cdot, \boldsymbol{x})$ of function $f(x)$. To calculate probabilistic extension, we use

$$\boldsymbol{f}(\xi, \boldsymbol{x}) = \sum_{i=1}^{n} \frac{\boldsymbol{x}(x_i(\xi))}{|f'(x_i(\xi))|}.$$

Example. Consider function $z = x + t$, where t is real number. Then probabilistic extensions can be represented as

$$\boldsymbol{f}(\xi, \boldsymbol{x}, t) = \boldsymbol{x}(\xi - t).$$

If $z = tx$, and $t > 0$, then probabilistic extensions can be represented as

$$\boldsymbol{f}(\xi, \boldsymbol{x}, t) = \frac{1}{t}\boldsymbol{x}(\frac{\xi}{t}).$$

Theorem 1.[7] Let $\boldsymbol{f}(\cdot, \boldsymbol{x}_1, \boldsymbol{x}_2, \dots, \boldsymbol{x}_n)$ be Probabilistic Extensions of function $f(x_1, x_2, \dots, x_n)$ and for each real t function $\boldsymbol{f}(\cdot, t, \boldsymbol{x}_2, \dots, \boldsymbol{x}_n)$ be probabilistic extensions of the function $f(t, x_2, \dots, x_n)$. Then

$$\boldsymbol{f}(z, \boldsymbol{x}_1, \boldsymbol{x}_2, \dots, \boldsymbol{x}_n) = \int \boldsymbol{x}_1(t) \boldsymbol{f}(z, t, \boldsymbol{x}_2, \dots, \boldsymbol{x}_n) dt \quad (1)$$

Corollary. Theorem 1 infers the possibility of recursive computations for the general form of probability extensions and reduction to the calculation of the one-dimensional case.

Let us consider the computing the integral (1). For simplicity, we represent (1) as a quadrature

$$\int \boldsymbol{x}_1(t)\boldsymbol{f}(z, t, \boldsymbol{x}_2, \ldots, \boldsymbol{x}_n)dt \approx \sum_{l=1}^{m} \gamma_l \boldsymbol{x}_1(t_l)\boldsymbol{f}(z, t_l, \boldsymbol{x}_2, \ldots, \boldsymbol{x}_n)$$

Further, we can also use numerical quadratures for computing $\boldsymbol{f}(z, t_l, \boldsymbol{x}_2, \ldots, \boldsymbol{x}_n)$. In general, it is NP-hard problem with actual parallelization.

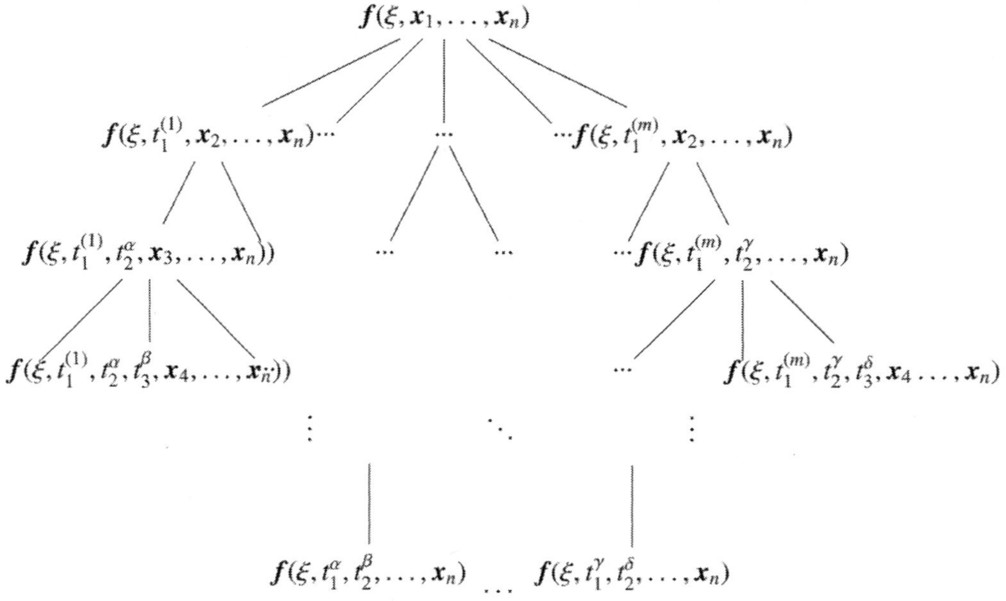

Figure 1. The tree of parallel recursive programming

In Figure 1 shows the tree of parallel recursive organization of the computational process. Thus, on the lower layer, it is necessary to compute the probabilistic extensions only for one variable. Note that all computations on each layer are independent and can be computed simultaneously.

Let x, y be system of two continuous random variables with probability density functions $\boldsymbol{x}, \boldsymbol{y}$.

Densities arising as a result of arithmetic operations on random variables have the form

$$(\boldsymbol{x} + \boldsymbol{y})(\xi) = \int_{\underline{x}}^{\overline{x}} \boldsymbol{x}(t)\boldsymbol{y}(\xi - t)dt;$$

if $\mathrm{supp}(\boldsymbol{x}) = [\underline{x}, \overline{x}]$ and $\underline{x} > 0$ then

$$(\boldsymbol{x}\boldsymbol{y})(\xi) = \int_{\underline{x}}^{\overline{x}} \frac{1}{t}\boldsymbol{x}(t)\boldsymbol{y}(\frac{\xi}{t})dt,$$

$$(\boldsymbol{y}/\boldsymbol{x})(\xi) = \int_{\underline{x}}^{\overline{x}} t\boldsymbol{x}(t)\boldsymbol{y}(t\xi)dt.$$

These arithmetic operations completely coincide with the classical operations on the probability density functions.

Let $f(x_1, \ldots, x_n)$ be a rational function. To construct a probabilistic extensions of f, we replace the arithmetic operations by the arithmetic operation on random variables, while the variables x_1, x_2, $\ldots$, x_n are replaced by probability density functions. It makes sense to call the resulting probabilistic extensions of f as *natural probabilistic extension.*

Case 1. [7] Let $x_1, \ldots, x_n$ be independent random variables. If $f(x_1, \ldots, x_n)$ is a rational expression where each variable x_i occurs not more than once, then the natural probabilistic extension approximates a probabilistic extension.

Case 2. [7] Let the function $f(x_1, \ldots, x_n)$ can be a change of variables, so that $f(z_1, \ldots, z_k)$ is a rational function of the variables $z_1, \ldots, z_k$ satisfying the conditions of Case 1. The variable z_i is a function of x_i, $i \in Ind_i$. and Ind_i be mutually disjoint. Suppose for each z_i is possible to construct a probabilistic extension. Then the natural extension $\boldsymbol{f}(z_1, \ldots, z_k)$ would be approximated by a probabilistic extension $\boldsymbol{f}(x_1, \ldots, x_n)$.

Example. Let $f(x_1, x_2) = (-x_1^2 + x_1)\sin x_2$, and $z_1 = (-x_1^2 + x_1)$, $z_2 = \sin x_2$. Notice that it is possible to construct probabilistic extensions for the functions z_1, z_2, and then compute $f = z_1 * z_2$, which is a rational function satisfying the conditions of Case 1. So, natural extension will approximate probabilistic extension for the function $f(x_1, x_2)$.

3 Systems of linear algebraic equations

As one example of numerical simulation, let us consider solution of a system of linear algebraic equations

$$Ax = b, \tag{2}$$

where $A = (a_{ij})$ a random matrix and $b = (\boldsymbol{b}_i)$ a random right-hand side vector respectively. Suppose that the random matrix A and the vector b have independent components with probability densities $\boldsymbol{A} = (\boldsymbol{a}_{ij})$, $\boldsymbol{b} = (\boldsymbol{b}_i)$ respectively and

$$\boldsymbol{A} = \begin{pmatrix} \boldsymbol{a}_{11} & \boldsymbol{a}_{12} & \ldots & \boldsymbol{a}_{1n} \\ \vdots & \vdots & \ddots & \vdots \\ \boldsymbol{a}_{n1} & \boldsymbol{a}_{n2} & \ldots & \boldsymbol{a}_{nn} \end{pmatrix}.$$

The support of the solution set can be represented as follows [6]

$$X = \{x | Ax = b, A \in \text{supp}(\boldsymbol{A}), b \in \text{supp}(\boldsymbol{b})\}.$$

Construct the probabilistic extension of the solution vector $\boldsymbol{x}(\cdot, \boldsymbol{A}, \boldsymbol{b})$

$$\boldsymbol{x}_1(\cdot, \boldsymbol{A}, \boldsymbol{b}) = \frac{\begin{vmatrix} \boldsymbol{b}_1 & \boldsymbol{a}_{12} & \ldots & \boldsymbol{a}_{1n} \\ \vdots & \vdots & \ddots & \vdots \\ \boldsymbol{b}_n & \boldsymbol{a}_{n2} & \ldots & \boldsymbol{a}_{nn} \end{vmatrix}}{\begin{vmatrix} \boldsymbol{a}_{11} & \boldsymbol{a}_{12} & \ldots & \boldsymbol{a}_{1n} \\ \vdots & \vdots & \ddots & \vdots \\ \boldsymbol{a}_{n1} & \boldsymbol{a}_{n2} & \ldots & \boldsymbol{a}_{nn} \end{vmatrix}}$$

or

$$\boldsymbol{x}_1(\xi, \boldsymbol{A}, \boldsymbol{b}) = \iint \boldsymbol{a}_{12}(t_{12}) \ldots \boldsymbol{a}_{nn}(t_{nn}) \frac{\sum \boldsymbol{b}_i \Delta_i(t_{12}, \ldots, t_{nn})}{\sum \boldsymbol{a}_{1i} \Delta_i(t_{12}, \ldots, t_{nn})}(\xi) dt_{12} \ldots dt_{nn}, \tag{3}$$

where $\Delta_i(t_{12}, \ldots, t_{nn}) \in R$ are minors from the Cramer method for solving SLAE, $t_{ij} \in \text{supp}(\boldsymbol{a}_{ij})$. The expression

$$\left(\frac{\sum \boldsymbol{b}_i \Delta_i(t_{12}, \ldots, t_{nn})}{\sum \boldsymbol{a}_{1i} \Delta_i(t_{12}, \ldots, t_{nn})} \right)(\xi)$$

is computed using probabilistic arithmetic.

Example. Consider a system of linear algebraic equations

$$Ax = b, \tag{4}$$

Let $A = (a_{ij})$ be random matrix $n = 2$. The elements of the matrix $\boldsymbol{A}$ are independent and distributed on the triangular law, $\boldsymbol{a}_{11}, \boldsymbol{a}_{22}$ distributed on the interval $[2, 4]$, a_{21}, a_{12} distributed on the interval $[-1, 1]$. The vector $\boldsymbol{b}$ consists of independent components $\boldsymbol{b}_1$, $\boldsymbol{b}_2$, they are distributed according to a triangular law on the interval $[0, 2]$.

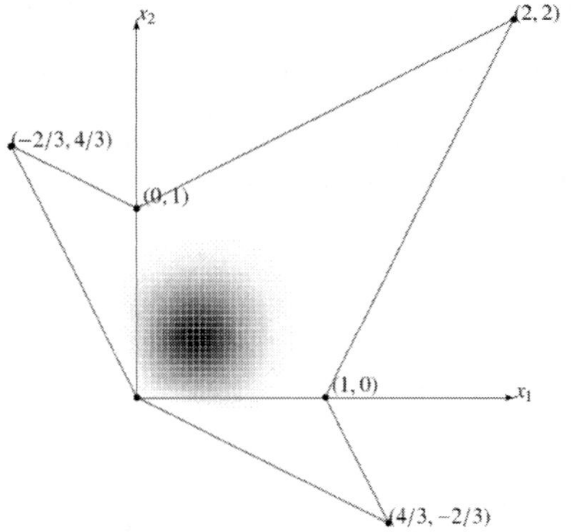

Figure 2. The boundary of the solution set and the joint probability density function of (x_1, x_2)

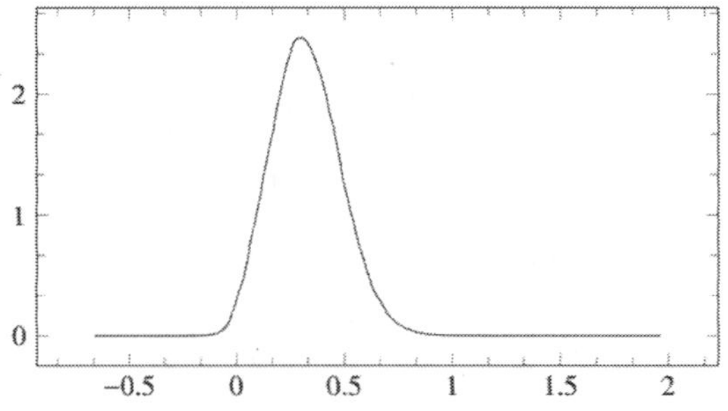

Figure 3. The probability density function of random variable x_1

The vector (x_1, x_2) solution of (4)

$$x_1 = \frac{a_{22}b_1 - a_{12}b_2}{a_{11}a_{22} - a_{12}a_{21}},$$

$$x_2 = \frac{a_{11}b_2 - a_{21}b_1}{a_{11}a_{22} - a_{12}a_{21}}.$$

Figure 2 shows the joint probability density of the vectors (x_1, x_2). The value of the probability is represented by shades of gray. A solid line is the boundary of the solution set.

In this case, to compute $\boldsymbol{x}_1$, we replace random variables a_{22}, a_{12} to the real t_{22}, t_{12}. Computing the integral numerically for different ξ

$$\boldsymbol{x}_1(\xi) = \iint \boldsymbol{a}_{22}(t_{22})\boldsymbol{a}_{12}(t_{12})\left(\frac{t_{22}\boldsymbol{b}_1 - t_{12}\boldsymbol{b}_2}{\boldsymbol{a}_{11}t_{22} - t_{12}\boldsymbol{a}_{21}}\right)(\xi)dt_{22}dt_{12}.$$

we obtain probability density function of x_1.

Figure 3 shows the probability density function of random variable x_1. The random variable x_1 is distributed on interval $[-2/3, 2]$. However, significant values are on the interval $[0, 1]$.

4 Conclusion

The proposed approach makes it possible to solve the problem of computing the probability density function in the modeling processes with random input data. For these purposes we propose using a parallel-recursive organization of the computational process. Thus, the important problem of computing probability extensions can be solved within parallel recursive programming. This opens multifold possibilities for studying issues of Earth remote sensing with random input data. Fast and accurate calculations are based on the properties of numerical arithmetic procedures over piecewise polynomial models developed within the framework of numerical probabilistic analysis.

References

[1] Hamideh Ebrahimi, Shadi Aslebagh, W. Linwood Jones Proceedings of IEEE Southeastcon (2013). DOI: 10.1109/SECON.2013.6567506

[2] Dobronets B.S., Popova O.A. Journal of Siberian Federal University. Engineering & Technologies, **9**(7), 960–971, (2016)

[3] Dobronets B.S., Popova O.A. Journal of Siberian Federal University, Mathematics and Physics, **10** (1), 16–21 (2017).

[4] Dobronets B.S., Popova O.A. IOP Conf. Series: Journal of Physics: Conf. Series **1015** (2018) 032028 doi :10.1088/1742-6596/1015/3/032028

[5] Dobronets B.S., Popova O.A. IOP Conf. Series: Materials Science and Engineering **354** (2018) 012006 doi:10.1088/1757-899X/354/1/012006

[6] Dobronets B., Popova O. *Scientific Computing, Computer Arithmetic, and Validated Numerics. Lecture Notes in Computer Science* **9553**. Springer International Publishing, 43–53 (2016)

[7] Dobronets B.S., Popova O.A. Reliable Computing. **19**, 274–289 (2014).

Multi-version approach to improve the reliability of processing data of the earth remote sensing in the real-time

Mikhail V. Saramud[1], *Igor V.* Kovalev[1,2], *Vasiliy V.* Losev[1], *Mariam O.* Petrosyan[1], *and Dmitriy I.* Kovalev[2]

[1] Reshetnev Siberian State University of Science and Technology 31, Krasnoyarsky Rabochy Av.,Russia, 660037
[2] Federal State Autonomous Educational Institution of Higher Education Siberian Federal University, 79, Svobodny Pr., Krasnoyarsk, Russia, 660041

Abstract. The article describes the use of a multi-version approach to improve the accuracy of the classification of images when solving the problem of image analysis for Earth remote sensing. The implementation of this approach makes it possible to reduce the classification error and, consequently, to increase the reliability of processing remote sensing data. A practical study was carried out in a multi-version real-time execution environment, which makes it possible to organize image processing on board of an unmanned vehicle. The results confirm the effectiveness of the proposed approach.

Krasnoyarsk region is a vast territory, which includes forests, requiring constant monitoring and control [1]. There is a need to quickly identify forest fires, pest infestations, illegal logging and other situations that may pose a danger to people and nature [2]. Earth remote sensing (ERS) allows solving this problem quickly, as well as conducting research, creating geographic information systems that are necessary for enterprises to solve production problems - from laying cable networks to the development of mineral deposits.

With Earth remote sensing, the task of automated processing of a huge amount of information arises. It is not possible to process such volumes manually using operators. Investigation of Earth remote sensing images by man is only possible for a specific area of interest due to its identification by an automated system or for other reasons. Modern image processing algorithms and decision-making algorithms based on them are not perfect. As a rule, the automated study of large areas is the task of dividing regions into given classes: “norm”, “fire”, “struck by a pest”, “cutting down”, etc [3]. However, classification errors are not excluded. In case of an erroneous classification of a normal area as a problem, for example, a “fire”, this will only lead to a waste of time for manual checking and additional research of this area. The case of the erroneous classification of the problem area as normal can lead to disastrous consequences. Thus, the problem arises of increasing the accuracy of the classification when processing images of Earth remote sensing.

One of the most effective known ways to improve the quality of software systems is a multi-version approach. To be able to use it, we need several different, but functionally equivalent software “versions” [4]. In our case, the versions correspond to different classification algorithms. In this task, multi-version can be easily achieved, since today there are many different algorithms of this class.

Moreover, there are often several sources of information about the same territory. These can be images in different ranges (infrared, ultraviolet, radio spectrum of the electromagnetic spectrum, etc.), from different satellites and aircrafts. In this case, you can create a multi-version pool of software implementations of a single algorithm that decides on various sources of information.

As practice shows, multi-version systems significantly increase both the reliability and the quality of decision making in the system. For practical testing of this approach in a real multi-version execution environment, three verification algorithms are implemented, each of which decides whether the same area belongs to one of the 6 specified classes. Based on the outputs of these algorithms, the system will make the final decision.

The software implementation of the environment in which the verification algorithms are executed is a multi-version real-time execution environment. This environment is based on the real-time operating system FreeRTOS v10 and is a set of C source files for compilation [5]. This allows you to make modifications and extensions to the code of the execution environment itself and to develop new modules.

The execution environment itself has rather low hardware requirements and allows assembly for most modern single-chip platforms. This fact allows organizing image processing directly "on board" of an autonomous air vehicle or space vehicle. In such conditions, the bottleneck is the communication channel, which does not allow transmitting a constant stream of images for processing to the ground. The ability to process, and most importantly, make decisions on board and transfer to earth only the decision made and only the image of the required area of interest, greatly simplifies the process of Earth remote sensing in order to quickly identify problem areas.

As a decision-making algorithm in multi-version systems in this software implementation, the author's modification of the weighted voting algorithm by a coordinated majority with the forgetting element is used [6]. The version weights provided for by this algorithm (necessary for the operation of the algorithm) are, in their sense, an assessment of the quality of the version, and therefore the algorithm it implements. The presence of version weights will not only increase the reliability at the decision-making stage, but also provide estimates of the quality of the algorithms used. Thus, we will receive additional information on the operation of the algorithms in real conditions of the execution environment.

Figure 1 shows the results of the operation of testing algorithms in a multi-version execution environment. Considering the results obtained, we note that the checking algorithms made 4, 5 and 9 errors, respectively.

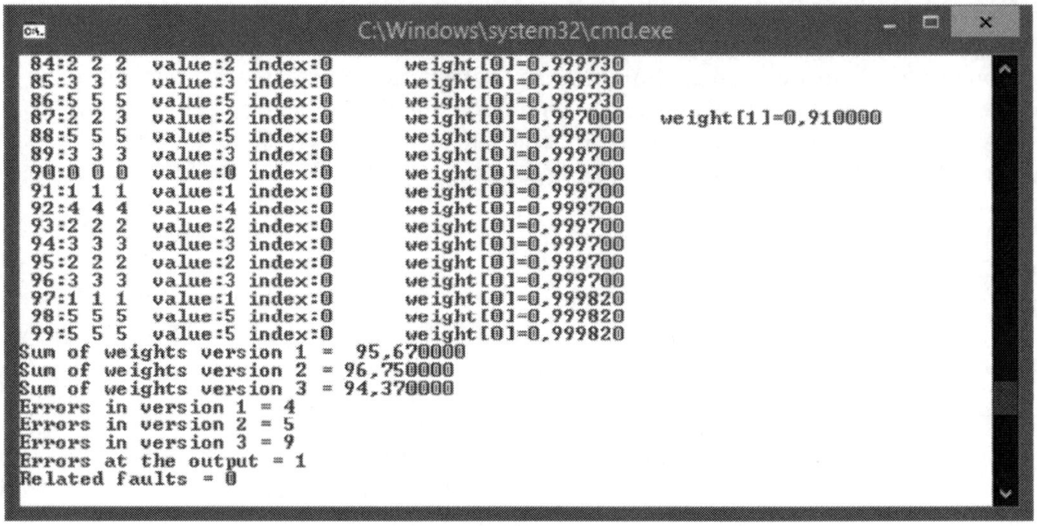

Fig. 1. The results of the work of verification algorithms in the execution environment.

Based on the responses of these algorithms, the multi-version system always chose the right answer, making only one mistake. That is, the system made four times less classification mistakes than the best algorithm. This proves the effectiveness of the multi-version approach, since the quality of the system operation with its application exceeds the quality of operation of any of the algorithms used in it.

Conclusion

The obtained results prove the applicability of the multi-version approach for solving the problem of increasing the reliability of processing remote sensing data and show the high efficiency of this approach. The use of multi-version models in data processing will increase the reliability of the results obtained and reduce the classification error.

The considered practical solution based on the real-time operating system FreeRTOS is fault tolerant, not demanding on the resources of the hardware platform and portable. This allows organizing a streaming data processing system, even within an unmanned vehicle. This is a very important factor, since it is extremely difficult to transmit a stream of images via communication channels during the flight, and if they are processed "on board," the device will already report the result - the identified problem in a certain area.

Such an opportunity will significantly speed up the detection of forest fires and other events requiring prompt response.

Acknowledgments

This work was supported by Ministry of Education and Science of Russian Federation within limits of state contract № 2.2867.2017/4.6.

References

1. Korobko A.V., Penkova T.G., Nicheporchuk V.V., Mihalev A.S. The integral OLAP-model of the emergency risk estimation in the case of Krasnoyarsk region // 2013 36th International Convention on Information and Communication Technology, Electronics and Microelectronics (MIPRO), pp. 1161 – 1166 (2013)

2. Quan Zou, Guoqing Li, Wenyang Yu An integrated disaster rapid cloud service platform using remote sensing data // 2017 IEEE International Geoscience and Remote Sensing Symposium (IGARSS), pp. 5221 – 5224 (2017)
3. Yuhuan Ren, Yalan Liu Geological disaster detection from remote sensing image based on experts' knowledge and image features // 2016 IEEE International Geoscience and Remote Sensing Symposium (IGARSS), pp. 677 – 680 (2016)
4. Kovalev I. V., Losev V. V., Saramud M.V., Kovalev D.I., Petrosyan M.O. To the question of implementation of multi-version execution environment software of onboard autonomous pilotless objects by means of real-time operating system // Vestnik SibGAU, 2017. Vol 18, №1, p. 58 – 61. (2017)
5. Saramud M.V., Kovalev I. V., Losev V. V., Petrosyan M.O. Application of FreeRTOS for implementation of the execution environment of real-time multi-version software // International Journal on Information Technologies & Security, No 3 (vol. 10), pp. 75 – 82. (2018)
6. Igor Kovalev, Anna Voroshilova, Vasiliy Losev, Mikhail Saramud, Maria Chuvashova, Aleksandr Medvedev. Comparative Tests of Decision Making Algorithms for a Multiversion Execution Environment of the Fault Tolerance Software // Proceedings of 2017 European Conference on Electrical Engineering and Computer Science (EECS 2017), pp 211 – 217 (2017)

The Technology to Identify Firebreak Plowing Objects Based on the Satellite Data of the Earth Remote Sensing

Evgenii A.Maltsev[1*], *Yuri* A. Maglinets[2], *Gennady* M.Tsibulskii[2]

[1]Skolkovo Institute of Science and Technology, Nobel Street 3, Moscow, 143026, Russia
[2]Siberian Federal University, Borisova street 26, Krasnoyarsk, 660074 Russia

Abstract. This work describes the technology to identify firebreak plowing objects of agriculture fields based on the satellite data of the Earth Remote Sensing within the medium and high spatial resolution. The technology uses a model of the firebreak plowing object, vegetation indexes and spatial relations between objects.

1 Introduction

The risk of a fire hazard emergence on agriculture fields increases in the cereals harvesting season due to straw and unplowed lands.

Therefore, it is necessary to plow the boundaries of agriculture fields in order to reduce fire hazard risk [1]. This condition sets a task of the control over compliance with fire safety regulations.

This paper shows the use of information model in the technology of firebreak plowing objects (FPO) identification.

This technology is a set of methods and algorithms that allow recognizing the target objects based on satellite data of the Earth Remote Sensing at medium and high spatial resolution. Besides, it allows identifying the territories requiring control over existing or missing FPO according to [1].

2 Fire Plowing Model

Firebreak plowing objects on digital images of the Earth Remote Sensing (ERS) within medium and high spatial resolution are objects of open soil with a linear geometric structure. The length of FPO varies depending on the perimeter of plowed agriculture fields. Width of FPO should be at least 4 meters [1]. There are time constraints for receiving the necessary remote sensing data because the period of cereals harvesting in autumn season is 1-2 months. The FPO areshowed in Fig. 1 (a-c).

* Corresponding author: evgenii.maltsev@gmail.com

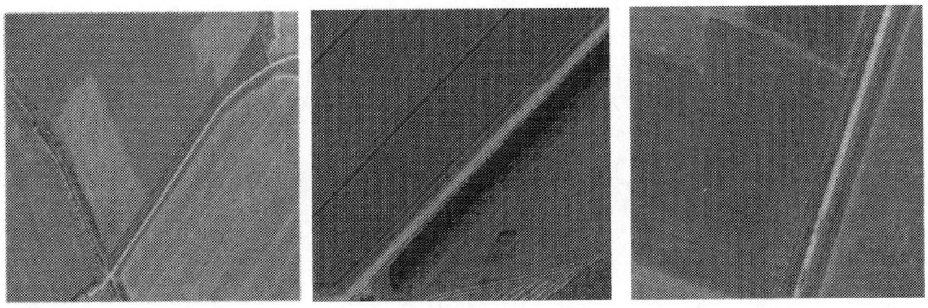

a *b* *c*

Fig. 1. Fragments of ERS data (SPOT 6) obtained for different areas within Krasnoyarsk Territory. FPO are highlighted in red color.

To solve the FPO identification task described in [2], the use of consolidated stages of the ERS digital image processing was investigated as well as methods of supervised classification, such as the method of parallelepipeds; minimum distance classification method; maximum likelihood classification method.

Applying the consolidated stages of ERS image interpretation produces a sufficient number of type 1 classification errors (false positive) because the spectral specifications of target objects have values of spectral specifications of open soil.

A fragment of a segmented satellite image is showed in Fig 2.

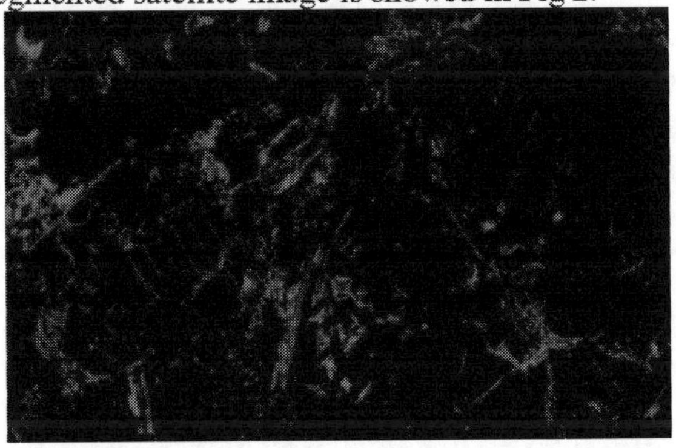

Fig. 2. A fragment of processed satellite image (segmented objects – red color, background – black color).

Thus, to solve the FPO identification task an image needs to be transformed from iconic processing level to relational structures level [3, 4] with calculation of metric, topological, spectral and spatial features of object, and also with creation of relations between objects.

A model of FPO as an identification object is described in [5].

FPO is an object with linear geometric structure [6] described by the following expression (1):

$$FPO = <M, T, L, N, E> \tag{1}$$

where M is spatial multiple connection; T are constraints of object width; L are constraints of object length; N is range of NDVI values corresponding to open soil; E are spatial relations between FPO and neighbouring objects.

The following expression is used for calculation of NDVI for a satellite image (2):

$$NDVI = \frac{NIR-RED}{NIR+RED} \tag{2}$$

where *NDVI* is a value of normalized differential vegetation index of satellite image pixel; *NIR* is a value of the nearest infrared spectral channel (wavelength 0.76-1 micrometres); *RED* is a value of red spectral channel (wavelength 0.6-0.76 micrometres);

To solve the task of open soil segmentation it is required to define LP (ω) predicate based on the spectral feature (NDVI value). NDVI values for the open soil are within the range [7,8]:

$$0.025 \leq \text{NDVI} \leq 0.26 \qquad (3)$$

Expression for LP (ω):

$$\text{LP}(\omega) = \begin{cases} \text{TRUE, if } 0.025 \leq \text{NDVI}(f(x,y)) \leq 0.26 \\ \text{FALSE, otherwise.} \end{cases} \qquad (4)$$

where *f(x,y)* – function of brightness of satellite image pixel; *NDVI* – value of normalized differential vegetation index of satellite image point.

The rule for FPO classification is as follows: the selection of segments belongs to the set of located territories using predicate LP, moreover, the thickness of segments is greater than the specified threshold and the length is even more greater than thickness:

$$\bar{g}(\omega) = T(\omega) > c_1 \; AND \; \omega \in W_1 \; AND \; T(\omega) \ll L(\omega) \; AND \; LP(\omega) \qquad (5)$$

where $T(\omega)$ – function of object thickness; $L(\omega)$ – function of object length; c_1 – set threshold of object thickness (4 meters for FPO); W_1 – multitude of located territories areas. Euclidean distance is the metrics of the classifier.

3 Interpretation of Agriculture Fields

One of the main tasks of agriculture field interpretation is determination of a set of features that separate agricultural fields from territories with natural vegetation.

The paper describes the differences between natural vegetation and several agriculture crops described in paper [9], such as duration of vegetation period, and interannual dynamics of green biomass. The vegetation index dynamic is used to estimate the start date of phenological phases of plants. Maximum vegetation index is suggested to correspond to blooming phase.

Paper [10] describes 3 periods of vegetation seasons when spectral differences between natural vegetation and spring crops are maximal:

1: (May - first half of June) Minimal coefficients of reflectance, period of the sowing;

2: (end of July – beginning of August) Maximal coefficients of reflectance of grain crops in spectral channel with wavelength 0.841-0.876 micrometres, period of heading and blooming.

3: (end of August – September) Maximal coefficients of reflectance in spectral channels with wavelength 0.620-0.670 and 0.841-0.876 micrometres, period of the full wax ripeness and start of harvest.

The analysis shows that NDVI and perpendicular vegetation index (PVI) are used in most of algorithms for agriculture fields interpretation with methods of supervised classification [11,12,13].

4 Technology of FPO Identification

The technology based on the general approach of pattern recognition and image analysis [14]

The technology includes the following stages [15]:

- The selection of the input ERS data with account for cloudiness is performed based on the developed regional center of ERS [16] using antenna receiving complex;
- Preliminary processing of the input ERS data till 2A level and also georeference using reference points, atmospheric correction and estimation of cloudiness. The latter estimation is completed using the algorithm described in [17].
- interpretation of objects of the agriculture fields,forests, automobile roads and railways, power lines [1].
- Location of territories for further data processing;
- Segmentation of open soil objects;
- Classification of FPO pursuant with (5);
- Estimation of classification accuracy;

5 Experimental Research over Technology of FPO Identification

Pursuant to main stages of the technology the expert selected the input satellite ERS data with the value of cloudiness being 5% using the catalogue of the regional ERS center data (Fig. 1). The catalogue is available on the Internet: http://digitalatlas.ru/. The next step of the technology is preliminary data processing.

Interpretation of additional objects (forests, automobile roads, railways, power lines) has been performed by the expert with ground-based researches and GPS binding.

The resulted vector layers are marked on Fig. 3. The scale of vector data is 1:10000. Metrical georeferencing error is 4 meters.

a) b) c)

Fig. 3. Fragments of ERS data with classified objects (automobile roads – white color; agriculture fields – yellow color; forest – green color)

The interpretation of additional objects was followed by location of territories areas to search FPO objects. For the latter purpose vector layers have been loaded to developed program module to locate FPO objects, territories have been selected for followed interpretation after processing (Fig. 4).

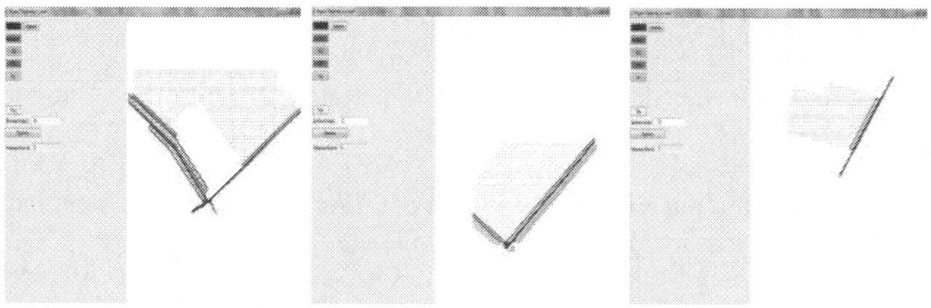

a) b) c)

Fig. 4. Operation of the module for location of the FPO with selected territory areas for further data interpretation.

The next step of the technology is segmentation of open soil using the calculated NDVI for satellite images and classification of FPO objects (Fig. 5).

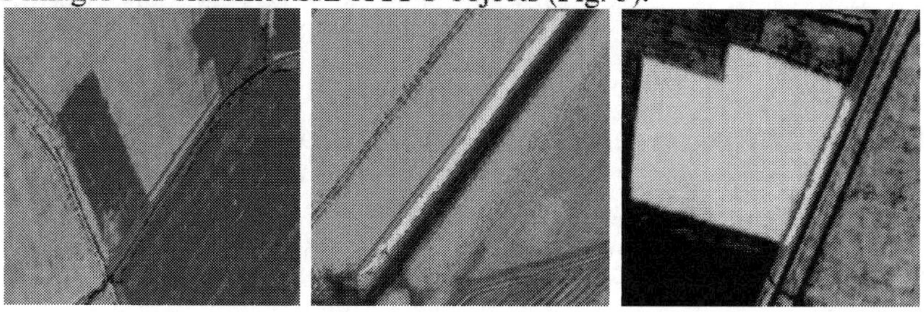

a) b) c)

Fig. 5. Images of the NDVI calculated on the basis of the satellite images with located territory areas for further interpretation (red color) and classified FPO object (blue color).

6 Conclusion

The technology of FPO identification has been developed using information model described in [5].

Over 100 satellite images with high spatial resolution picturing the agricultural areas within Krasnoyarsk Territory have been processed using the proposed technology. The resulted classification accuracy is 93%.

In addition, task solving allowed developing ancillary algorithms, methods, and software to identify FPO. Experimental research has been completed over the proposed technology within the agricultural areas of Krasnoyarsk Territory.

The reported study was funded by Russian Foundation for Basic Research (No.18-47-242002), Government of Krasnoyarsk Territory, Krasnoyarsk Regional Fund of Science, to the research project: «The development of technology for creating intelligent information systems of object-oriented monitoring of territories based on remote sensing data».

References

1. Order No 313 issued by the Ministry of the Russian Federation for Civil Defense, Emergencies and Elimination of Consequences of Natural Disasters on June 18 "On approval of rules fire safety in the Russian Federation(2003).
2. A.M. Chandra, S.K. Gosh, Remote sensing and geographic information systems, 312 p,(2008).

3. G.M. Tsibulsky, Y.A. Maglinets, Problems technic and technology XXI centure, About semiotic approach for creation expert systems, pp.195-202 (1994).
4. G.M. Tsibulsky, Informational problems of region, Symbol systems and search in analyze tasks and recognition images (1998).
5. E.A. Maltsev, Regional problems of the Earth Remote Sensing, Development of information model of the firebreak plowing objects for the automated processing of satellite data, pp. 326-329 (2014)
6. E.A. Maltsev, PhD dissertation, Development of the Earth remote sensing data processing algorithms for recognition of spatial agriculture objects with linear geometric structure, (2017)
7. [Electronic resource] Geographical information systems and remote sensing. - Access: http://gis-lab.info/qa/ndvi.html
8. R.V. Brejnev, Y.A. Maglinets, K.V. Raevich, G.M. Tsibulsky,ITNT, The modeling spatial agriculture object with heterogeneous dynamic variable spatial structure, pp 1090-1099 (2018)
9. I.A. Neishtadt, S.A. Bartalev, E.A. Loupian ,Current problems in remote sensing of the Earth from space, The approach for detection of used plowable land by satellite remote sensing,**3**, pp. 271-280 (2006)
10. Y.M. Sultangazin, Current problems in remote sensing of the Earth from space, ,Monitoring of the rotation of arable land in Northern Kazakhstan according to TERRA / MODIS data, **2**,pp. 302-307 (2005)
11. I.A. Neishtadt, S.A. Bartalev,D.M. Ershov, E.A. Loupian, I.A. Savin, Geoinformatics, Algorithms for analyzing the data of TERRA-MODIS satellite observations for monitoring agricultural land, pp. 205-209 (2004)
12. S.A. Bartalev, E.A. Loupian, I.A. Neishtadt, , I.A. Savin, Exploring the Earth from Space, Classification of some types of agricultural crops in southern regions of Russia according to MODIS satellite data,pp. 68-75 (2006)
13. Y.M. Sultangazin, N.R.Muratova,A.G. Terehov, Current problems in remote sensing of the Earth from space, The use of space monitoring in the planning and forecasting of grain production parameters, **1**, pp. 291-297 (2004)
14. Denisov, D.A. Computer methods of video information analysis [Text]: monograph / D.A. Denisov; Krasnoyarsk State Technical University – Krasnoyarsk, 1993. -192 pages.
15. Maltsev E.A. Development of the Technology to Identify Firebreak Plowing Objects Based on the Satellite Data of the Earth Remote Sensing./ E.A. Maltsev // Journal of Siberian Federal University. Engineering & Technologies, 2017, 10(6), 812-818
16. Maglinets Y.A. Multipurpose Geoinformation Management System of Territories Along the Yenisei Meridian/ Maglinets Y.A., Tsibulsky G.M., Maltsev E.A.// Pattern Recognition and Image Analysis. Advances in Mathematical Theory and Applications", том. 21, 2011,No. 2, стр. 290–292 с.]
17. Maltsev E.A., Sirotin E.Ye, Perfilyev D.A., Tsibulsky G.M. "Extracting a Cluster of Clouds in the SPOT 4 Satellite Images" [Text] // International Journal of "Pattern Recognition and Image Analysis. Advances in Mathematical Theory and Applications", vol. 21, 2011, No. 2, pp. 501-504.

Forest vegetation Interaction with L-band Satellite Signals

Vladimir Podoprigora[1,2] , and *Anatoly Sorokin*[2,3]

[1]Siberian Federal University, 79 Svobodny av., 660041 Krasnoyarsk, Russia.
[2]Federal Research Center, 50 Akademgorodok, 6660036 Krasnoyarsk, Russia
[3]Siberian University of Science and Technology, 31 Krasnoyarskij Rabochij av., 660037 Krasnoyarsk, Russia

Abstract. A method for calculating the electro-physical parameters of forest vegetation cover interacting with the signals of navigation satellites in the frequency bands 1.2-1.6 GHz is proposed. A technique for the experimental measurement of linear attenuation coefficient of satellite signals on test areas of canopies was developed, taking into account their calibration at the input. The proposed forest vegetation model describes the group of crowns as a system of quasi spherical "capsules", located chaotically in the air matrix. The scattering and penetration cross sections and the coefficient of linear attenuation of radio waves were calculated using the approximation of anomalous diffraction for crowns and the results of solving the diffraction problem for trunks. The effects of multiple scattering were taken into account.

1 Introduction

Currently, the Earth's forest vegetation cover monitoring using modern technologies in remote sensing over high frequency (UHF) radio waves becomes more actual. To determine the state of forested area using remote radio wave satellite sounding, it is necessary to solve the inverse problem based on the finding functional and statistical dependencies between the electro physical parameters of the forest and the received satellite signals. However, it is possible to found the relation between biometric and taxation characteristics of the forest vegetation cover and parameters of the processed signals only based on the corresponding electro physical models of the forest vegetation.

2 Model of the interaction the L-band signals with the forest

To describe the propagation of radio emission in the forest, the models [1] are used. They represent the vegetation cover either as a continuous dielectric medium with random or as a discrete set of elements. The application a particular model is specified by the conditions of small dimensions of the forest elements (leaves, needles, branches, trunks) in comparison with the wavelength of the incoming radiation. In the microwave range of satellite signals, the dimensions of forest elements are commensurable with the wavelength λ of an electromagnetic wave (λ = 1-30 cm), so the calculation of scattering, penetration and attenuation cross sections should be based on diffraction models that take into account the shape and dimensions of the elements. Rigorous solution of the diffraction problem for disks, strips (leaves), cylinders (needles, trunks, branches) does not exist. Therefore, these sections can be found only under certain restrictions. There are not enough works analyzing the relation between satellite radiofrequencies in the L-band (1.2-1.6 GHz) and the forest area, moreover their conclusions can be ambiguous.

The present study is devoted to the development of techniques and experimental measurement of the signals of GLONASS and GPS satellites interacting with forest vegetation coverage of various types, as well as methods for calculating the basic parameters of this interaction: the coefficient of linear attenuation, scattering and penetration cross sections.

Three areas of the Krasnoyarsk Academgorodok with an area of ≈ 900-1000 m^2 each were used as the experimental site was used by there. It were covered with pine and birch trees with different thicknesses, without undergrowth. The age of trees was 60-70 years. The average size parameters are given in Table 1.

Table 1. Forest stand parameters

Forest type	Height of trees, m	Crown height, m	Diameter of the trunk, m	Cross-sectional area, m^2
Pine	23	7	0.31	0.075
Birch	21	8	0.32	0.080

The average distance between trees for a dense tree stand was 3.5 m. The density of trees was 0.11 trunks/m^2 (1100 trees per 1 hectare), the average distance between the pines for broken forest was 9.9 m with the density 0.05 trunks/m^2. The average distance between the birches was 8.2 m, the density was 0.06 trunks/m^2(600 trees per 1 hectare).

The effective dielectric constant at a frequency of 1.4 GHz was chosen: for coniferous forest E = 1.001 + 0.0002i; for deciduous forest E = 1.0028 + 0.0006i [2].

The processes of scattering, penetration and attenuation in the most general discrete model are described by the classical Rayleigh-Gans (RG) transformation. Point dipoles are often used for particles with dimensions much smaller than the length of an electromagnetic wave. Large scatterers are described by the anomalous diffraction (AD). Other models reviewed in [1], for example, the radiation transfer model [3], MIMIC [4], are also used.

This paper represents a forest model that describes a group of crowns as a system of quasi-spherical “capsules” with radius R. They are located randomly in the air matrix and have a different internal structure of the branches arrangement depending on the tree species. Since crowns in decimeter wave range have a value of $kR = 2\pi R/\lambda >> 1$ at R = 1.5-2 m and the relative refractive index of a crown is close to one, the AD approximation was used to calculate the scattering cross section.

In this approximation, assuming the crown of the tree to be a "capsule" with a radial arrangement of branches, we obtain we obtain total cross section the scattering σ_p for the one tree crown for the total cross section:

$$\sigma_p = 2\sigma_0 \int_0^1 (2 - cos\Delta_{ll} - cos\Delta_l) y dy. \quad (1)$$

Here σ_0 is a geometric cross-sections of scattering, $\Delta_{//}$ *and* $\Delta_{/}$ are quantities that depend on the effective permittivity of the tree crown, its radius R and the angle between the wave vector $\bar{\boldsymbol{k}}$ and the vertical line; y is a integration variable equal to the relation to the radius of scattering vector projection onto a plane orthogonal to the wave propagation vector [5].

The effects of multiple scattering on spherical tree crowns were taken into account in calculating the differential scattering cross-section $(\frac{d\sigma_p}{d\Omega})_N$ by a system of N spheres, which can be written as

$$(\frac{d\sigma_p}{d\Omega})_N = N\left\{\left(\frac{d\sigma_p}{d\Omega}\right)_N\left[1 + N_v g(k_p, \eta, R)\right]\right\} = N\frac{d\sigma_p^{эф}}{d\Omega}, \qquad (2)$$

where Ω is a solid angle of scattering, N_v is an average density of spheres,

$$g(k, \eta, R) = 4\pi\int_0^{\infty} G(r^1)\frac{\sin k_p r^1}{k_p r^1} r^2 dr^1 \qquad (3)$$

is the Fourier transform of the radial distribution function $G(r^1)$, which depends on the difference between the wave vectors of incoming and scattered waves $k_p = k - k^1$, a radius of the spheres R and the volume η occupied by spheres of crowns, $\eta = N_v(\frac{4}{3}\pi R^3)$; $\frac{d\sigma_p^{эф}}{d\Omega}$ is the effective differential scattering cross-section. The equation (2) consists of two parts: 1) the scattering differential cross-section of an isolated sphere multiplied by *N*; 2) the correlation member $N_v g(k_p, \eta, R)\frac{d\sigma_p}{d\Omega}N$, which depends on the phase difference between the scattered fields and the position of the spheres in the ensemble.

The most realistic way of taking into account the correlation effects in scattering is the correlation function used in [6] and suitable for describing media with high concentration of particles.

The dependences $d\sigma_p/d\Omega$ on the scattering angle calculated from formula (2) at different tree concentrations η show that the effects of multiple radio waves scattering on crowns are noticeably manifested at small observation angles. The influence of crowns concentration on the degree of scattering medium homogeneity can be characterized by the ratio $\sigma_p^{эф}/\sigma_p$.

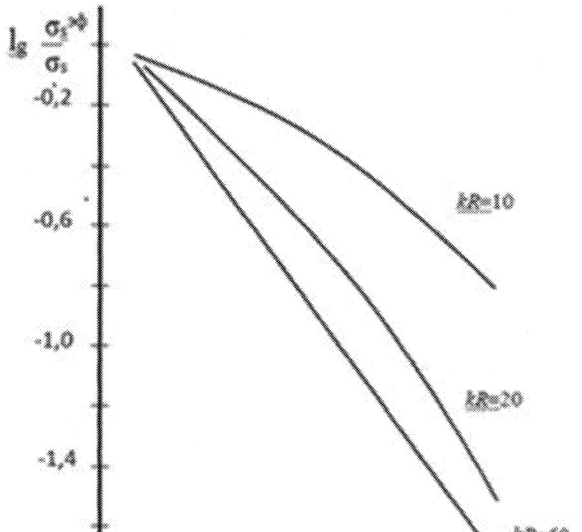

Fig.1. Dependencies of $\log\frac{\sigma_p^{эф}}{\sigma_p}$ on the volume concentration of crowns η for different values of *kR*.

It can be seen from Fig. 1., that as the concentration increases, the effective scattering cross-section $\sigma_p^{эф}$ decreases in comparison with σ_p.

The different slope of the curves $f(\eta) = \frac{\sigma_p^{эф}}{\sigma_p}$ indicates that the degree of homogeneity depends on the size of the spheres. The radiation transfer equations showed that at small concentration of particles in the medium $\sigma_p^{э\phi}$ is equal to the scattering cross-section of a particular particle in the matrix. When concentration increases, a value of $\sigma_p^{э\phi}$decreases, and at $\sigma_p^{э\phi} \ll \sigma_p$ the "crown-air" scattering medium can be considered as quasi homogeneous, therewith the radio wave propagation in it can be described as in the case of continuous homogeneous medium.

The attenuation and the scattering cross-sections [7, 8] at the trunks and branches were calculated using the equations derived from the solution to the diffraction problem for infinite circular cylinders. They was presented as series containing the functions of Bessel and Hankel. The calculation of the cross-sections σ_0 *and* σ_p shows that the effect of resonance interaction of L- band radio waves with trunks and branches is significant. The group of the trunks are presented by a statistical ensemble characterized by a normal distribution of tree parameters eight, diameter and distance between them). The average orientation of the thick branches, contributing to the scattering cross-section, is described by the "order parameter", determined by the slope angles cosines of the branches relative to the vertical.

3 Conclusion

The experimental studies were carried out using the method for determining the local coefficient of linear attenuation of γ signals from navigation satellites in the forest developed in [9]. The values of γ, measured in a variation of electromagnetic transmission probing for pine and birch stands, are in satisfactory agreement with the calculated values of the linear attenuation coefficient. For example, for the GLONASS satellite system and the pine forest $\gamma_{exp} = -0{,}26 \pm 0{,}02\ dB/m$, $\gamma_{calculated} = -0{,}23\ dB/m$. This /study allows us to hope for the applicability of the described approach for solving other problems of remote sensing of forest vegetation.

References

1. V.I. Popov, *Propagation of radio waves in forests* - M.: Hot line - Telecom,.- 392 p. (2015)
2. A.A. Milshin,A.G.Granov. Study of the Earth from Space.**3**, 50 (2000).
3. A.Ishimaru. *Wave Pro pagation and Sca tt ering in Random Media* (Academic Press, New York,1978).
4. F. Ulaby, K.Sarabandy, Mc Donald Int. J. Remote Sensing. **11**, 7, 1223 (1990).
5. S.ZumerPhys. Rev., **37,** 10, 4006 (1988) .
6. H.Eyring, D. Henderson, B. Stover,*Statis fecal Mechanies and Dynamics* - Wiley, New York, 1982) .
7. A.Chukhlantsev, A. Shutko, S. Golovachev. Radio engineering and electronics, **48**, 11, 1285 (2003).
8. S. Novik, A.Zavyalov, E. Telpuhovsky Proceedings of higher educational institutions. Physics. **3**, 57 (2006).
9. A. Sorokin, V. Podoprigora, S. Fomin, I Savin. Proceedings of higher educational institutions. Physics. **60**, 12, 19 (2017).

Dynamic texture recognition under adverse lighting and weather conditions for outdoor environments

Anna Pyataeva[1,*]

[1]Siberian Federal University, Institute of Space and Information Technology, 26, Kirensky st., Krasnoyarsk, 660074, Russian Federation

Abstract. Recognizing dynamic patterns based on visual processing is significant for many applications. In this paper dynamic texture recognition focuses on outdoor scenarios where a crisis event might occur (i.e. fire in a forest, floods/flooding etc.) Real outdoor scenes may include the objects with dynamic behaviour due to illumination, blurring, or weather conditions effects. Under bad weather conditions the imaging systems is degraded and produce low visibility images. In this work precipitation artefacts and lightning effects for dynamic texture analysis were studied. Experimental results show that the proposed method of weather and adverse lighting effects compensation is feasible and effective for video-based dynamic texture analysis under bad weather conditions.

1 Introduction

Nowadays dynamic textures (DT) recognition plays an important role in different computer vision community tasks in a variety of fields such as urban and forest scenes. Most of the existing methods in the field of computer vision are based on the assumption that the weather in outdoor images or videos is clear. Real outdoor scenes may include the objects with dynamic behaviour because of possible varying illumination, blurring, or weather conditions. Under bad weather conditions the imaging systems is degraded and produce low visibility images. Such effects may significantly degrade the performance of outdoor vision systems which relies on image/video. The removal of these effects provides accuracy increasing of video based computer vision algorithms.

The goals of the dynamic textures recognition can be different. The dynamic scenes analysis in video is a very useful task especially for the detection and monitoring of natural hazards such as floods and fires. The analysis of such videos is considered of utmost importance during natural disasters, since it can improve situational awareness by providing early detection of floods and fires to prevent or reduce damage from such emergencies.

In this paper we target the problem of classifying such DTs as water, smoke and flame.

2 Related works

* Corresponding author: anna4u@list.ru

Video based detection algorithms rate significantly depends of the varying illumination or weather conditions. One of the most important factors is illumination level. The influence of shadows and time of day on performance of video detection systems at signalized intersections can provide from 2 to 36% false alert errors [1]. Lighting variations are a challenge in face recognition. In paper [2], focusing on the illumination change challenge, a deep network model is proposed which takes into account both visible light image and near-infrared image to perform face recognition. Decreasing of computer vision algotytms accuracy due to illumination changes is shown for various tasks [3,4]. Rain and snow detection algorithms are usually classified into three categories: optical approaches [5, 6], spatial-domain approaches [7], and frequency-domain approaches [8]. Usually rain and snow compensation method consists of two stages: precipitation detection and precipitation removal from video images. For precipitation artifacts detection Kim et al proposed a method based on a Gaussian distribution model and block-matching algorithm [9]. Weather detection stage may extract weather cues such as sky, shadow, reflection, contrast and haze [10, 11]. Shen et al label snow in videos and extract the alpha value of the snow particles by alpha matting and general image contrast [12]. As one of the state-of-the-art motion-based rain removal algorithms, Chen et al. proposed a rain-pixel recovery algorithm based on motion segmentation of dynamic scene [13].

3 Detecting and extracting precipitation and lightning artifacts

In this section, the most suitable approaches for decetion and extraction weather and lightning artefacts are presented. One may refer the weather effects (Section 3.1) and adverse lightning (Section 3.2).

3.1 Weather conditions

There are the following precipitation types: haze, fog, rain and snow. Weather effects due to precipitations presistance in outdoor scenes may gives various effects in video images accoding to its type and intensity. For example, a rainy scene is typically composed of many raindrops – each with a size much larger than visible wave lengths, which causes the light to spread in a specific way. Closer raindrops, which have a projected size larger than a pixel, are individually visible, and further ones can only be seen in a macroscopic view as participating media [14]. Weather scenes include several effects such as streaks due to the velocity of raindrops and camera's exposure time, individual snow or rain particles and visibility is decreasing until details blur into a fog like appearance. To compensate the weather effects at the first step, the block-matching algorithm for each block is performed between adjacent frames in order to find blocks similar to the block that has weather artifact pixels. Size of blocks is 15-30 pixels with search radius defended in 5-20 pixels according to rain and snow particles speed. For moving blocks HOG – features of rain and snow are estimated. In weather compensation stage the video images were prossesed by VirtualDub, a free and open-source video-capture and video-processing utility for Microsoft Windows written by Avery Lee. Temporal Smoother, 2D Cleaner, MSU Denoiser, Smart Smoother, MSU Smart Sharpen, Sharpen filters were used for weather artifacts removal. The proposed removal algorithm is based on many filters, so it is very efficient for particles and tracks deleting. After filters application Dark Channel prior was used for haze removal. Dark Channel prior is based on atmospheric scattering model [15]. So haze-free image is recovered by:

$$J(x,y) = \frac{I_C(x,y) \cdot map(x,y)}{A_C \cdot (1 - map(x,y))}, \tag{1}$$

where $I_C(x, y)$ is the observed intensity in pixel with coordinates (x, y) in RGB-channel C, A_C is the global atmospheric light in each of RBG channel, $map(x, y)$ is the medium transmission describing the portion of the light that is not scattered and reaches the camera computed by:

$$map(x, y) = 1 - \phi \min_{C \in \{r,g,b\}} (\min_{y \in \Omega(x)} \left(\frac{I_C(x, y)}{A_C} \right), \qquad (2)$$

where φ=0,75 is the scattering coefficient of the atmosphere in pixel (x, y), $\Omega(x, y)$ is a local patch centered at pixel (x, y). Finally, median filtering operator is performed by Eq. 3 for visibility enhancement and flickering artifacts removal.

$$G_{i,j} = med[E_{i+s,j+t}; (s,t) \in W]; \qquad i, j \in Z^2. \qquad (3)$$

The above precipitation effects removal algorithm, has achieved its high efficiency for snow, rain, haze and other weather artifacts, deleting them from video images.

3.2 Adverse lightning compensation

The adverse image histogram shown in Fig. 1 provides identification of adverse lightning images properties such as: most of the pixels have small intensity values, intensity range constriction and low contrast.

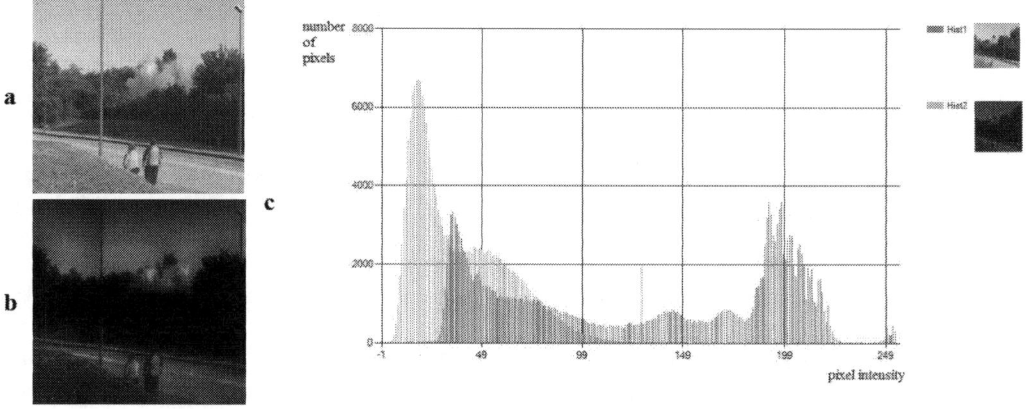

Fig. 1. (a) original image; (b) dark image; (c) histograms of images.

For illumination effects compensation and visual quality enhancement images need to average pixel intensity increase, expanding the range of brightness, increasing image contrast and eliminating influence of the additive noise. The proposed algorithm employs Multi Scale Retinex with Color Restoration (MSRCR) calculated using Eq. 3 [16].

$$R_{MSRCR_i}(x, y) = C_i(x, y) \cdot R_{MSR_i}(x, y), \qquad (3)$$

where $C_i(x, y)$ is the color restoration function for RGB color channel numbered i. Color restoration function is computed according to Eq. 4.

$$C_i(x, y) = \beta \log[\alpha_i I'(x, y)], i \in R, G, B \qquad (4)$$

In Eq. 4 α is a gain constant, and β controls the strength of the non-linearity constant, i is a color channel value. $I'(x, y)$ is given below:

$$I_i'(x, y) = I_i(x, y) / \sum_{i=1}^{S} I_i(x, y), \qquad (5)$$

where $I_i(x, y)$ is the pixel intensity with (x, y) coordinates. Mathematical representation of MSRCR (3) R_{SRi} describes computing of Multi Scale Retinex by Eq. 6:

$$R_{MSRi}(x,y)=\sum_{n=1}^{N}\omega_n \log\left[\frac{I_i(x,y)}{F_n(x,y)*I_i(x,y)}\right], \quad (6)$$

where N is the number of scales, ω_n is the weight of each scale, $F_n(x, y)$ is the Gaussian surround function, symbol * denotes convolution. After the application of MSRCR algorithm Laplacian, Gaussian and median filters for image enhancement were used.

4 DT recognition algorithm

4.1 Dynamic textures

In the context of surveillance, recognizing dynamic patterns is of significance to isolate activities of interest (e.g. fire) from distracting background (e.g. windblown vegetation and changes in scene illumination). Dynamic textures can be defined as visual phenomena that exhibit spatial and temporal regularity. A complex approach for early detection of floods and fires by prevent or reduce the damage from such emergencies primarily requires the detection of such categories of DT like flame, smoke and water.

4.2 DT recognition algorithm

In detail DT recognition algorithm was considered by the authors at previous work [17]. The first step of the proposed algorithm is motion estimation based on Block-matching SAD criteria and the optical flow for the information about the local and global motion vectors. Block-matching algorithm evolves 2 to 5 frames of video sequences according to experiments for objects in different Categories show various motion speed. The next algorithm step is colour, facial, entropy features estimation. The fourth step is Laws energy approach for edges, spots, ripples and wave texture features detection. The next step is ELBP descriptors computing. Then histogram approach was applied. Chi-square distance, histogram intersection distance, Kullback-Leibler divergence, and G-statistic are usually used during classification stage. In this research, the histogram intersection was chosen for histogram comparison. Regions clustering based on boosted random forests.

5 Experimentation

For experimentation results Dyntex dataset [18] and VMOTE dataset [19] with such DT objects like smoke, flame and water were used. Examples of video frames are depicted in Fig. 2. In this video sequences precipitation and lightning effects were added artificially. For adverse lightning imitation in this work Gamma correction, additive Gaussian noise and impulse noise were applied successively. Precipitation images imitation was generated using Gaussian filter.

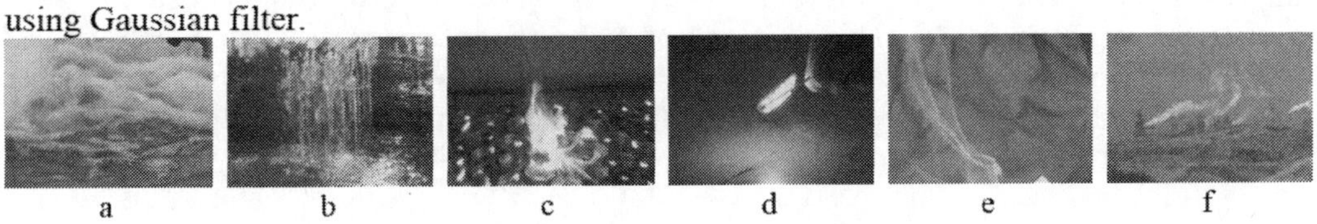

Fig. 2. (a)-(b) water; (c)-(d) flame; (e)-(f) smoke.

Video sequences with natural snow, rain and adverse lighting from the real surveillance cameras were downloaded from YouTube. In addition, the representativeness of the test set is increased by a video sequence, which the authors themselves recorded at night. Compensation algorithm experimental results are shown in Fig. 3.

Fig. 3. Compensation algorithm experimental results: a - weather effects; b – artifacts is removed

In table 1 DT detection results for videos with flame, smoke and water were demonstrated. The performance of the DT classification algorithm was evaluated using the CR, % – classification rate, FRR, % – false rate rejection and FAR, % - false alert rejection.

Table 1. DT detection results.

Video (DT type)	Original video			Weather effects			Artifacts is removed		
	CR	FAR	FRR	CR	FAR	FRR	CR	FAR	FRR
VMOTE\DV004 (smoke)	95,5	4,53	2,12	75,5	24,5	24	94,16	5,84	2,00
Dyntex\649g910.avi (smoke)	100	0,00	3,12	60,2	39,8	18,45	99,23	0,77	2,98
Dyntex\648dc10.avi (water)	99,0	1,99	2,42	65,2	34,8	15,45	97,45	2,55	2,07
Dyntex\6489610.avi (water)	90,4	9,57	3,07	58,2	41,8	30,4	89,21	10,79	3,00
Dyntex\6ammj00.avi (flame)	96,0	0,00	3,27	82,4	17,6	29,45	95,21	4,79	3,14
Dyntex\64cac10.avi (flame)	92,4	7,87	4,00	80,2	19,8	17,42	90,52	9,48	2,54
Averaged values	95,55	3,99	3,00	70,28	29,72	22,53	94,3	5,70	2,62

True recognition for smoke in adverse lighting conditions is degraded to 70%. False rare rejection and false alert errors significantly increase to 29.7% and 22.5% respectively. After adverse lighting and weather effects compensation algorithm work true recognition of DT regions increases to 94,3%. This accuracy provides the influence of adverse lighting and weather effects on a quality of DT detection is studied. Experimental results show that the proposed method is feasible and effective for video-based dynamic texture analysis in varying illumination and weather conditions.

6 CONCLUSIONS

Real outdoor scenes may include the objects with dynamic behavior because of possible varying illumination, blurring, or weather conditions. Due to weather and lightning effects persistence averaged accuracy of DT recognition is decreasing to 70%. The minimum detection accuracy is 58 %. To compensate weather effect for damaged blocks the video images were processed by VirtualDub using a series of filters to suppress additive noise, blur, sharpness. After filters employment Dark Channel prior was used for haze removal. For illumination effects compensation and visual quality enhancement images the proposed algorithm employs MSRCR, Laplacian, Gaussian and median filters. This work focuses on detecting of dynamic textures in video sequences captured by surveillance cameras. In the aspect of preventing natural disasters, the work is focused on the detection of such types of DT as water, fire and smoke. After adverse lighting and weather effects compensation algorithm work true recognition of DT regions increases to 94,3%. This accuracy provides the influence of adverse lighting and weather effects on a quality of DT detection is studied. Experimental results show that the proposed method is feasible and effective for video-based dynamic texture analysis under bad weather conditions.

References

1. M. V. Chitturi, J. C. Medina. Transportation Research, **18(2),** 176-186 (2010).
2. K. Guo, S. Wu, Y. Xu. CAAI Trans. on Intelligence Technology, **2(1)**, 39-47 (2017).
3. R. Ramamoorthi. IEEE Trans. on Pattern Analysis and Machine Intelligence, **24(10)**, 10-21 (2002).
4. T. Riklin-Raviv, A. Shashua. Proceedings of the Conf. on Comp. Vis. and Pattern R., 566-571 (1999).
5. K. Garg, S. K. Nayar. Conf. on Comp. Vis. and Pattern R., **1,** 528-535 (2004).
6. X. Zhang, H. Li, Y. Qi, W. K. Leow. Int. Conf. Multimedia Expo. **1,** 461-464 (2006).
7. J. H. Kim, C. Lee, J. Y. Sim, C. S. Kim. Conf. Image Processing. **1**, 914-917 (2013).
8. P. Barnum, T. Kanade, S.G. Narasimhan. Int. Conf. on Comp. Vis., (2007).
9. H. G. Kim, S. J. Seo, B. C. Song. J. Vis. Commun. Image R., 317-328 (2015).
10. Z. Zheng, M. Huadong, H. Fu, C. Zhang. Neurocomputing. 207, 365-373 (2016).
11. L. Cewu, L. Di, J. Jiaya. Conf. on Comp. Vis. and Pattern R, 3718-3725 (2014).
12. Y. Shen, L.Z. Ma, H. Liu, Y. Bao, Z. Chen. Inf. Processing Letters, 1124-1130 (2010).
13. J. Chen, L.P. Chau. IEEE Trans. on Image Process. **23 (3),** 1097–1104 (2014).
14. Y.Weber, V. Jolivet, G. Gilet, D. Ghazanfarpour. Comp. & Graphics. **50,** 61-70 (2015).
15. K. He, J. Sun, X. Tang. The Proc. of Conf. on Comp. Vis. and Pattern R. 1956-1963 (2009).
16. S. Parthasarathy, P. Sankaran. National Conf. on Communications (2012).
17. A. Pyataeva, K. Raevich. Int. J. of Advanced Studies. **2.** 58-67 (2018).
18. P. Renaud, S.Fazekas, M. J. Huiskes. Pattern R. L. **31.** 1627-1632 (2010).
19. VMOTE Dataset. http://www2.imse-cnm.csic.es/vmote/english_version/index.php.

Semi-supervised learning through hierarchical clustering for interactive aerospace image analysis

Sergey Rylov[1,*]

[1]Institute of Computational Technologies of Siberian Branch of the Russian Academy of Sciences, 630090, Academician M.A. Lavrentiev avenue, 6, Novosibirsk, Russia

Abstract. A new semi-supervised classification algorithm based on the non-parametric clustering algorithm HCA is proposed. The algorithm obtains hierarchical segmentation result where additional classes that are not represented in the training samples can be found. High performance of the algorithm allows using it in interactive mode. Experimental studies confirm that the proposed algorithm provides aerospace image classification in conditions of limited number of training samples.

1 Introduction

In many practical problems of data classification, such as satellite images, the process of collecting qualitative training samples (TS) is usually connected with considerable material and time expenditure [1-4]. Therefore, in practice training samples are often limited and unpresentable. Besides, TS can be missing for some classes causing an adverse effect on the classification quality. The use of insufficient training samples does not provide satisfactory quality of segmentation, especially in the analysis of large and complex scenes.

In conditions of insufficient training samples, it is expedient to use classification methods with semi-supervised learning. In the course of their work, not only the information contained in the labelled training samples is used but also unlabelled (unclassified) data is utilized to construct the decision rule. The problem of semi-supervised classification can be considered in two formulations: as a classification that additionally uses information from unclassified data; and as clustering that uses additional conditions from training samples [1, 11]. The use of semi-supervised learning algorithms can reduce manual labour and improve the quality of classification results [3-4].

Over the past few years, works devoted to application of semi-supervised classification methods to satellite images have appeared [5-10], but their practical use is rather difficult due to the large number of configurable parameters and high computational costs [9]. Thus, it is urgent to develop new computationally efficient classification algorithms with semi-supervised learning that will allow processing aerospace images in conditions of small volume of training samples.

* Corresponding author: RylovS@mail.ru

This paper presents a new computationally efficient semi-supervised classification algorithm for aerospace image segmentation, based on the hierarchical grid-based clustering algorithm (HCA) [12], where training samples are taken into consideration at the hierarchy construction stage. The hierarchy itself is based on the set of small groups of data, obtained by nonparametric density estimation in the space of spectral features. The result is represented as a hierarchical segmentation, where, in addition to the required classes, there can be also found other classes not represented in the training samples. Algorithm's high performance provides the expert with the opportunity to use it in an interactive mode, setting extra training samples to get more accurate results if necessary. Experimental results confirm that the proposed algorithm allows classifying images in conditions of small size training samples.

2 Grid-based hierarchical clustering algorithm (HCA)

This section provides a brief description of the clustering algorithm HCA [12], which can be divided into four main stages.

At the first stage, a grid structure is formed in the feature space — it is divided into disjoint cells, the size of which depends on the parameter m. The density of a cell is determined through the number of data elements trapped in this cell.

At the second stage, each non-empty cell is connected to the corresponding neighbouring cell with the highest density. Thus, the set of non-empty cells is divided into connected components $\{G_1,\dots,G_S\}$, which correspond to single-mode clusters. In the HCA algorithm, we take the connected components as the basic elements of the hierarchy. The number of obtained connected components is small relative to the number of all data elements, and therefore the hierarchy construction does not require high computational costs.

At the third stage, the distances between all pairs of adjacent components are found. The distance is determined by estimating the density drop between the components. As a result, the corresponding distance matrix $\{h_{ij}\}$ is formed.

The distances between arbitrary connected components $\{\hat{h}_{ij}\}$ are determined through the distances between adjacent connected components $\{h_{ij}\}$ as follows. Let $\Theta_{ij} = \{Q_{ij}\}$ be the set of all chains of the connected components $Q_{ij} = \langle G_i = G_{k_1},\dots, G_{k_t}, G_{k_{t+1}},\dots, G_{k_l} = G_j \rangle$ such that for all $t = 1,\dots,l-1$, the components G_{k_t}, $G_{k_{t+1}}$ are adjacent. Then the distance between arbitrary connected components G_i and G_j is defined by the formula

$$\hat{h}_{ij} = \min_{Q_{ij}\in\Theta_{ij}} \left[\max{}_t\, h_{k_t,k_{t+1}}\right] \tag{1}$$

If the set Θ_{ij} is empty, we assume that $\hat{h}_{ij} = 1$.

The advantage of the introduced distance $\{\hat{h}_{ij}\}$ is that it has ultrametric property [13], i.e. it is a metric that satisfies the strong triangle inequality: $\hat{h}_{ij} \le max\,(\hat{h}_{ik}, \hat{h}_{kj})$, $\forall\, i, j, k$. It is known that there exists a unique correspondence between the distance matrices with ultrametric property and dendrograms [14]. Therefore, such matrix describes some hierarchical partition.

At the final stage, the single linkage algorithm (SLINK) for dendrogram construction is applied to the distance matrix $\{h_{ij}\}$. As a result, we obtain a dendrogram corresponding to the ultrametric $\{\hat{h}_{ij}\}$.

The clustering algorithm HCA allows obtaining hierarchical clustering structure of the data, while it is able to separate clusters intersecting in the feature space. The use of non-parametric density estimation allows distinguishing clusters of complex shape. The implemented algorithm is capable of processing data with up to 8 dimensions and up to several hundred millions of elements. At the same time, its computational efficiency allows clustering multispectral four-band images up to 100 million pixels in size within one second on a regular PC.

3 Semi-supervised hierarchy construction

In the proposed approach, the training samples are taken into account at the stage of hierarchy construction. When clustering with HCA algorithm, the hierarchy is formed with SLINK dendrogram construction method, which is applied to the distance matrix between adjacent connected components. The SLINK method for an $n \times n$ size matrix consists of (n-1) iterations. At each iteration, two nearest elements are united, while the distance to the new combined element is defined as the minimum of the distances to the elements to be united. There exists an algorithm [15] based on the use of an array of partial minima (containing indices and values of minimal elements in each row of the distance matrix), which allows finding the minimum matrix element at each iteration with $O(n)$ operations. Thus, the overall achieved computational complexity of the SLINK algorithm is $O(n^2)$.

Within semi-supervised approach, before the dendrogram construction, we establish classes for the connected components due to the available training samples. The component is assigned to the class of the training samples, which fall into this component. If the component contains training samples from different classes, then this component is marked as a `conflicting` one, and its class is established by voting of the TS elements. Components that do not contain a single TS element remain unlabelled. At this stage, the distances between the `non-conflicting` components belonging to the same class are set to zero. Thus, at the stage of hierarchy construction, such components are immediately united into one class.

Component's class labels are considered at the stage of dendrogram construction. At each iteration of the SLINK algorithm, the class labels of the elements to be united are checked. If they belong to different classes, then the union does not take place. Instead, the distance between them is set to one, and the minima in the corresponding rows of the matrix are updated. In other cases, the union takes place as usual. At the same time when a labelled element is united with an unlabelled one, the combined element inherits the class of the labelled element. Thus, the unlabelled components are eventually united with the closest labelled components in the metric (1). As a result, the top of the hierarchy consists of the combined components belonging to the required classes, as well as the groups of unlabelled components that turned out to be at distance 1 from all the marked ones (meaning they represent their own classes).

It is possible to retrieve data partitioning with various detail degree from the given hierarchical structure by varying the dendrogram cut value. Moreover, the classes known from training samples will stay separated at all levels. After obtaining the result, the user can interactively add new elements to the training samples, and the algorithm will only require performing anew the dendrogram construction step to update the result, which does not take significant time.

4 Experimental studies

In this section, the results of the proposed semi-supervised classification algorithm on model data and images are presented. It is shown that with the use of small size training samples (TS), the proposed approach allows distinguishing similar classes successfully, while avoiding unnecessary fragmentation.

Fig. 1,*a* represents the two-dimensional model dataset consisting of five classes with normal distribution, simulating vegetation classes (birch forests, coniferous forests, meadow vegetation, wetlands, agricultural land) on Landsat satellite image in red and green spectral features. Training samples (8 points) are marked on the image with black dots. With the use of these limited TS, classification algorithms make significant mistakes on the red class. For example, overall accuracy of Minimum Distance classification is 93,32 % (Fig. 1,*c*). At the same time, clustering algorithms have a problem to separate strongly intersecting purple and blue classes. Expectation-maximization clustering overall accuracy reaches only 89,28 % (Fig. 1,*d*). Nevertheless, the proposed semi-supervised classification algorithm managed to successfully extract all classes with 96,24 % overall accuracy (Fig. 1,*b*).

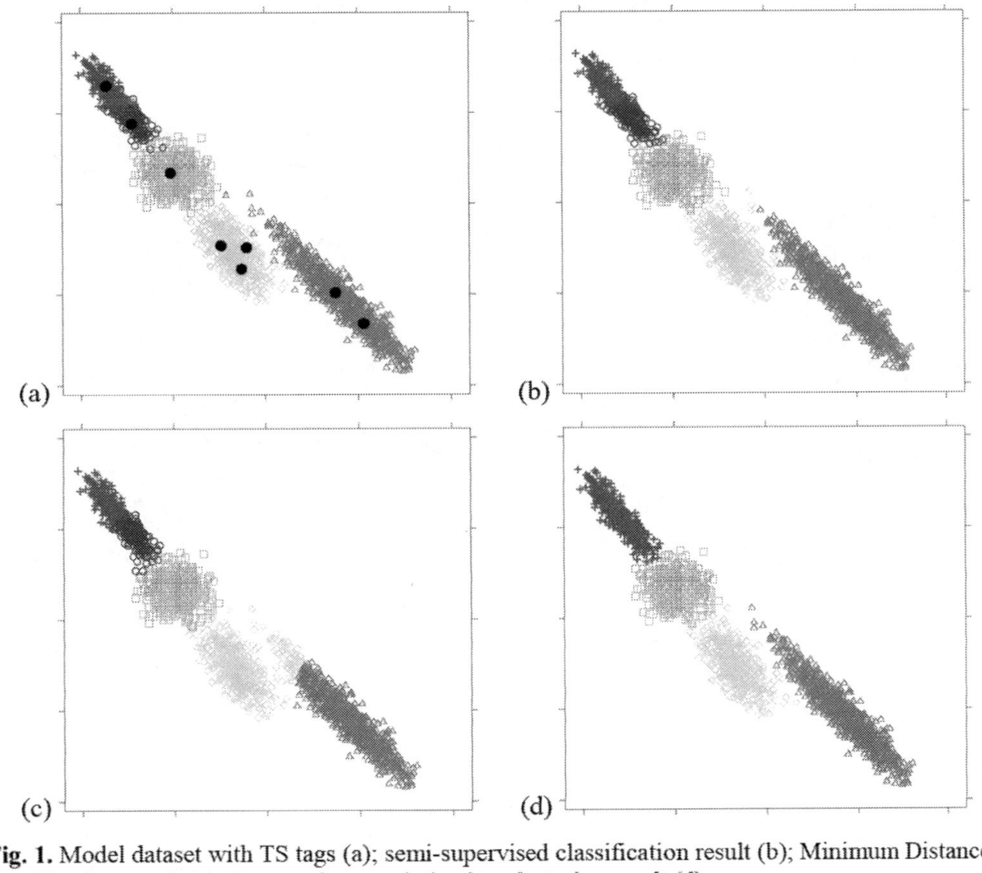

Fig. 1. Model dataset with TS tags (a); semi-supervised classification result (b); Minimum Distance classification result (c); Expectation-maximization clustering result (d)

Fig. 2 presents the model colour image containing the ring and the background, which consist of gradient transitions of various colours. Also, the image contains tags of the training samples (4 points for the ring and 5 points for the background). When clustering this image, a fragmented segmentation result is obtained, where both classes break up into many clusters, otherwise the ring and the background cannot be separated. However, with the given limited training samples, the proposed semi-supervised approach successfully distinguishes required classes (Fig. 2).

Fig. 3 shows the semi-supervised classification result of the high spatial resolution multispectral satellite image with training samples consisting of only 8 points. As a result, all 6 required classes were successfully extracted, including a forest that is extremely heterogeneous in its spectral characteristics. Moreover, the proposed algorithm additionally marked out a class that is not represented in the training samples, which corresponds to a football field (marked in white colour).

Fig. 2. Model image with TS tags (left); HCA algorithm clustering result (centre); semi-supervised classification result (right)

Fig. 3. WorldView-2 satellite image (RGB composite, bands 4, 6, 1) with TS tags and the result of the semi-supervised classification algorithm

5 Conclusion

The proposed semi-supervised classification algorithm can perform aerospace image segmentation under conditions of small and incomplete training samples. This provides the potential to the effective use of data obtained from field studies, which are commonly not used directly in the automated satellite image processing. Hierarchical representation of the segmentation result and high performance of the algorithm significantly facilitate the work of an expert.

For further research, we are planning to implement automatic partition of the `conflicting` components with a view to better fit with the training samples.

This work was supported by the Russian Foundation for Basic Research (No. 18-37-00492 mol_a).

References

1. O.I. Travkin, *Proc. XVIII Intern. Conf.: Analytics and data management in data-intensive areas*, 361-369 (Ershovo, Moscow Region, 2016)

2. Yu.V. Adaskina, A.M. Popov, P.V. Rebrova, *Proc. XVIII Intern. Conf.: IMS*, 15-24 (St. Petersburg, 2015)
3. V.J. Prakash, L.M. Nithya, IJCTT, **8**, 25-29 (2014)
4. M.F.A. Hady, F. Schwenker, *Handbook on Neural Information Processing*, 215-239. (Springer, 2013)
5. B. Banerjee, K.M. Buddhiraju, Journal of the Indian Society of Remote Sensing, **43 (4)**, 719-728 (2015)
6. K. Tan et al., ISPRS Journal of Photogrammetry and Remote Sensing, **97**, 36-45 (2014)
7. R. Luo et al., SPIE Remote Sensing, **10004**, 100040T (2016)
8. L. Yang et al., IEEE Geoscience and Remote Sensing Letters, **11 (3)**, 651-655 (2014)
9. L. Wang et al., ISPRS Journal of Photogrammetry and Remote Sensing, **97**, 123-137 (2014)
10. X. Jing, S.Y. Chen, L.L. Fan, *Proc. Ninth Intern. Conf.: ICDIP 2017*, **10420**, 1042030 (2017)
11. L. Lelis, J. Sander, *Proc. Ninth IEEE Intern. Conf. : ICDM'09*, 842-847 (IEEE, 2009)
12. S.A. Rylov, I.A. Pestunov, Journal of Physics: Conference Series (IOP Publishing, 2018) (to be published)
13. B. Leclerc, Math. Sci. Humaines, **127 (73)**, 5-37 (1981)
14. A. Mirzaei, M. Rahmati, IEEE Tr. Fuzzy Syst, **18 (1)**, 27-39 (2010)
15. C.F. Olson, Parallel computing, **21 (8)**, 1313-1325 (1995)

Extension of training set using mean shift procedure for aerospace images classification

Yuriy N. Sinyavskiy[1,*], *Pavel V.* Melnikov[1], and *Igor A.* Pestunov[1]

[1] Institute of Computational Technologies of the Siberian Branch of the Russian Academy of Sciences, Novosibirsk, Russia

Abstract. An effective method of training set extension for aerospace images classification is proposed. The method is based on mean shift procedure with respect to spatial information. It allows considering the unlabeled data structure. The results of experimental study using the Salinas hyperspectral image are presented, proving the effectiveness of the proposed method.

1. Introduction

In image classification tasks, a training set (labeled data) is necessary for constructing a decision rule or training neural networks [1, 2]. The process of obtaining a training set (TS) for aerospace image is often associated with significant material and time costs. Therefore, in practice, TS is available only for a small number of classes which is of interest to the user and is non-representative at the same time. Some classes can be represented by several labeled pixels.

It is known [3, 4] that in order to ensure an acceptable classification quality the minimum number of TS points for parametric classifiers should be of $10k$ per class order (where k is the dimensionality of the feature space), and for non-parametric classifiers should be of $50k$ per class. Therefore, the problem of obtaining a representative TS is especially relevant while processing hyperspectral images for which the number of spectral channels (features) is measured in hundreds.

At the same time, a characteristic feature of aerospace images classification tasks is that a large amount of unlabeled data is always available while solving them. In such circumstances the training set can be extended with the help of unlabeled data using methods based on clustering algorithms. Parametric methods based on the EM algorithm were most widely spread among such methods [5]. However, when processing aerospace images any *a priori* data on the probabilistic characteristics of classes are absent, as a rule. Therefore, the application of these methods can lead to unsatisfactory results.

A method of training set extension based on the nonparametric algorithm soft-PARZEN is proposed in [6, 7]. In this method "rigid" constraints on the form of conditional density function are not required. However, its application requires the presence of training sets for entirely all classes in the image. In practice, this condition is provided very rarely. In

* Corresponding author: yorikmail@gmail.com

addition, its application for processing aerospace images is associated with unacceptably high computational costs.

A common approach to the training set extension is based on the mean shift procedure. In [8] a semi-supervised support vector machine method is proposed. The support vector machine is applied to unlabeled data in such a way as to minimize an error when classifying both labeled and unlabeled data. The main disadvantage of the method is that the objective function for such a task is hard to optimize because it is not a convex one [9]. To solve this problem, a computationally effective gradient descent optimization method is suggested in [10].

In this work the method of training set extension based on mean shift segmentation with respect to the image spatial information is presented [11]. To demonstrate the efficiency of the proposed algorithm the constructed training sets were applied in order to classify the hyperspectral image of Salinas [12] with the Support Vector Machine (SVM) method using radial basis functions.

2. The proposed method of training set extension

The algorithm of the training set extension with the parameters $\{h_r, h_s, \alpha, p\}$ can be divided into three stages.

At the first stage, the image is segmented using the mean shift procedure with respect to spatial information. For this purpose mean shift procedure is applied to each pixel of the image. Only the pixels located at a distance of no more than h_r in the image domain and no more than h_s in the feature space [11] are taken into account when calculating the coordinates of a new center. The Euclidean distance between vectors is used as the distance between pixels. The distance is determined by the spectral brightness vectors in the feature space, and by the pixel coordinates vectors in the image domain. After that, segments with close centers (at a distance of no more than h_r in the image domain and no more than $h_s/2$ in the feature space) are combined and a representative is calculated for each segment (the average value of the feature vectors of all the pixels related to the segment).

At the second stage, a set of segments is formed for each class of the initial TS containing the points of the initial training set for this class. The segments from the obtained sets that simultaneously contain points from several classes of the initial TS are deleted. The appearance of such segments is possible in 2 cases. First, the parameters of the mean shift procedure can be unsuccessful. Second, errors can appear in the initial TS. Using the points from such segments in TS extension can lead to classification errors. Then, each set a extended by the segments located in the feature space at a distance of no more than α from it. The distance to a set is calculated as the smallest of the distances to its elements. The Euclidean distance between representatives is used as a measure of the distance between segments. After that, the segments that are simultaneously included in several sets are deleted again. These segments are located in the feature space on the boundary of the classes represented in the initial TS.

At the final stage of the algorithm, randomly selected points from the segments (p percent of the segment size) which are included in the set formed for this class are added to the initial TS for each class.

3. Experimental research

The proposed method of the TS extension is implemented in the C++ programming language (using Microsoft Visual Studio 2017 IDE) using the OpenMP standard.

A Salinas hyperspectral image (the Salinas Valley, California) measuring 512 × 217 pixels obtained from the AVIRIS sensor on October 8, 1998 was used in the experiment [12]. The features constructed by the principal component analysis were used for processing. Fig. 1 shows the image in false colors (the first three principal components) and the ground truth image for 16 classes.

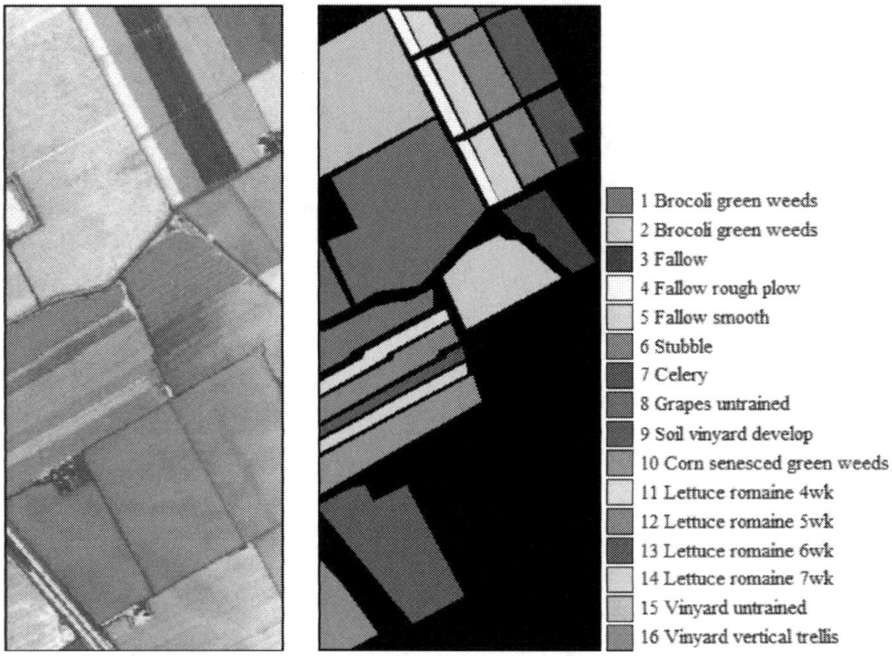

Fig. 1. Salinas image in false colors (first 3 principal components) and ground truth image for 16 classes.

The initial training sets containing 48 points (3 points per each class represented on the image), 80 points (5 points per class), and 160 points (10 points per class) were randomly formed according to the ground truth image. Each of these sets was extended by 5, 10, 15 and 20%. The parameter value $\alpha = 10$ was fixed. The extension of the sets was carried out using the first 4 principal components. The transition to the principal components is due to the high computational complexity of the mean shift procedure associated with the need for multiple calculation of the Euclidean distance in a multidimensional feature space. The resulting sets were used to classify the image using the SVM algorithm based on radial basis functions. The classification was carried out in the feature space obtained by the principal components analysis. The implementation of the SVM algorithm included in the Exelis ENVI software package was used. The default values were selected for all configurable parameters of the algorithm. The accuracy of the classification was defined by comparing the obtained classification images with ground truth one. Black pixels on ground truth image were ignored.

To average the results, the initial training sets were formed three times and for each of these sets the experiment with the same set of parameters was repeated three times. The obtained accuracy values of the classification (in percents) are given in the table. Fig. 2 shows the dependence of the average and maximum classification accuracy on the value of parameter p. The value $p = 0$ corresponds to the initial TS.

 https://doi.org/10.1051/e3sconf/20197501010

Accuracy values (in percents) for Salinas image classification with SVM algorithm using resulting TSs. The parameter value $p = 0$ corresponds to the initial TS.

Initial TS size	Value of parameter p												
	0	5			10			15			20		
48	12.0	66.2	64.1	65.0	67.5	68.6	68.1	70.6	69.0	69.6	69.9	69.5	69.5
	8.9	67.8	67.4	67.2	71.0	72.1	72.4	73.1	72.8	73.0	73.7	73.6	73.6
	9.5	60.7	61.2	60.6	61.5	61.7	61.7	62.8	62.5	62.6	64.3	63.7	62.9
80	33.2	73.9	74.9	73.9	77.0	76.6	76.7	77.2	77.7	76.4	77.4	78.1	77.6
	36.1	74.2	74.7	74.2	77.6	77.3	76.7	77.4	77.8	78.1	79.1	79.0	78.8
	39.8	64.0	64.0	65.7	67.4	67.3	66.2	67.8	68.1	67.1	67.9	67.5	68.3
160	51.2	62.9	62.3	61.7	63.6	63.7	63.1	63.8	63.6	63.4	64.0	63.3	63.6
	47.9	74.9	74.8	75.7	76.6	76.7	77.0	77.8	77.1	77.1	78.0	77.6	77.3
	53.8	66.4	67.3	67.5	70.5	70.9	71.1	72.2	72.2	75.5	73.2	73.5	73.5

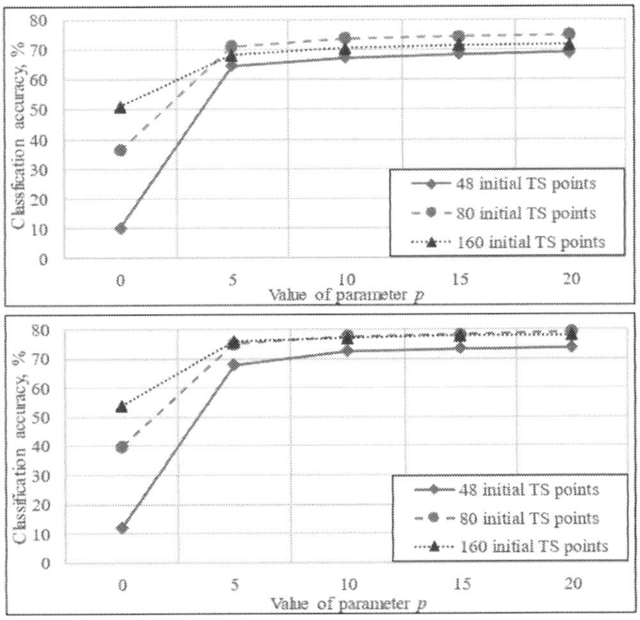

Fig. 2. The average (top) and maximum (bottom) accuracy of the Salinas image classification by SVM algorithm depending on the value of parameter p for different initial TS size. The value $p = 0$ corresponds to initial TS.

4. Conclusion

The work suggests an effective method of the training set extension for the tasks of aerospace images classification that allows to take into account the structure of unlabeled data. An experimental research using Salinas hyperspectral image showed that the proposed method is effective even for a small size of the initial set (3 points per class).

The work was carried out with partial financial support of the Integrated Program of Fundamental Research of the SB RAS "Interdisciplinary Integration Studies" for 2018-2020 (project No. 37) and the Russian Foundation for Basic Research (grant No. 18-37-00492).

References

1. A. Oliver, A. Odena, C. Raffel, E.D. Cubuk, I.J. Goodfellow, *Realistic evaluation of deep semi-supervised learning algorithms*, https://arxiv.org/pdf/1804.09170.pdf (2018)
2. Z. Dai, Z. Yang, F. Yang, W.W. Cohen, R.R. Salakhutdinov, Adv. Neural Inform. Processing Systems, 6510-6520 (2017)
3. Sh. Yu. Raudis, Statistical Problems of Control, **66**, 9-42 (Mathematics and Cybernetics Institute of LTS AS, Vilnius, 1984)
4. J.A. Richards, *Remote Sensing Digital Image Analysis* (Springer-Verlag, Berlin, 1999)
5. M.M. Dundar, D.A. Landgrebe, IEEE Trans. on Geosci. and Remote Sensing **42**, 264-270 (2004)
6. P. Juszczak, R.P.W. Duin, *Proc. 4th Intern. Conf. on Comp. Recogn. Syst.*, 203-210 (Springer-Verlag, 2005)
7. E.A. Kulikova, I.A. Pestunov, KazNU Bulletin. Mathematics, Mechanics, Computer Science Series, **13** (3), 284-290 (2008)
8. K.P. Bennett, A. Demiriz, Advances in Neural Information Processing Systems. 368-374 (1999)
9. X. Zhu, Comp. Sci. **2**, 4 (University of Wisconsin-Madison, 2006)
10. F. Gieseke, A. Airola, T. Pahikkala, O. Kramer, Neurocomputing **123**, 23-32 (2014)
11. D. Comaniciu, P. Meer, IEEE Trans. Patt. Anal. Mach. Intell. **24**, 603-619 (2002)
12. Hyperspectral Remote Sensing Scenes. http://www.ehu.eus/ccwintco/index.php/Hyperspectral_Remote_Sensing_Scenes#Salinas_scene (2014)

Forest dielectric characteristics for navigation satellites signals in L 1 range

Anatoly Sorokin[1,2,*] *Vladimir* Podoprigora[1,3], and *Evgenii* Ponomarev[1]

[1]Federal Research Center KSC SB RAS 50, Akademgorodok av., 660036, Krasnoyarsk, Russia
[2]Reshetnev Siberian State University of Science and Technology, Krasnoyarsky Rabochy Av., 660037, Krasnoyarsk, Russia
[3]Siberian Federal University, 79 Svobodny pr., 660041 Krasnoyarsk, Russia

Abstract. A structural dielectric model of the forest is presented taking into account the wood Biomass distribution. The multi-scale structural formations of the individual trees and the stand as a whole as a hierarchical system of non-interacting components of the mixed dielectric are discussed. Layers of crowns and trunks are distinguished as anisotropic dielectric media with gradient boundaries. The possibility of using the orientation parameter of the order of branches in the crown layer to estimate the gradients of the effective permittivity of the inhomogeneous distribution and its spatial anisotropy is analyzed. A comparative analysis of the features of the effective dielectric permittivity of pine and birch forests is carried out. The possibilities of using a layered anisotropic dielectric forest model to describe the transformation of GPS and GLONASS signals in order to restore the biometric characteristics of the forest are considered.

1 Introduction

The development of methods for recovering physical characteristics of tree stands using GLONASS and GPS signals has prospects of creating an effective forest monitoring system by a universal user resource - global navigation satellite systems (GNSS) [1]. Available GNSS signals are presented as an effective tool for continuous local monitoring of forests which allows obtaining data about the physical characteristics and state of the tree stands cover with a sampling frequency of 1 Hz in the L-band operating frequency range. Scanning of the stand with multidirectional microwave streams allows restoring the distribution of phytomass and the moisture of the stand of forested areas of up to 3 hectares on the vicinity of the antenna location. The duration of diurnal and seasonal processes of changes in water content in trees and air in the forest canopy compared to the duration of GNSS probing signals (1 ms) is essential. The hardware-software complex for recording changes in the characteristics of transmitted and scattered GNSS signals in the tree stands provides obtaining attenuation coefficients from localized areas of forests [2].

* Corresponding author: sorav@iph.krasn.ru

The presence of models of electromagnetic waves interaction of microwave range with a stand allows restoring quantitative values of taxation parameters and moisture content of forests from linear attenuation coefficients. The solution of the restoration task involves the development of adaptive models that take into account the structural and species diversity of individual trees and the stand as a whole and as a hierarchical system radio physical of components of a mixed dielectric.

In this work, a version of the development of a dielectric model of a stand adapted for the characteristics of GNSS signals of the L1-band is presented.

2 Special features of the dielectric forest model for GNSS signals

Tree stands are randomly inhomogeneous medium with tree elements of various shapes, sizes and spatial orientation [3-4]. A universal approach to describing the electro physical properties of different types of forests is possible when considering a forest as a multiphase dielectric medium with losses characterized by an effective dielectric permittivity (EDP). The forest as a two-component continuous dielectric medium, which includes multi-scale tree elements and air. It can be considered as a mixed inhomogeneous dielectric. Experimental data on the elements dependence of the complex permittivity tensor of wood $\varepsilon = \varepsilon' + i\,\varepsilon''$ of different types of forests on humidity and temperature for radio frequencies in the range of 20-1011 Hz are presented in the monograph [5]. The difference in dielectric properties of wood is observed for orientations of the electric field strength vector along the tree in the radial and tangential directions. In practical applications tensor elements corresponding to the longitudinal direction (*L*) and the cross one (*C*) are used. The anisotropy of wood is characterized by the value of $\Delta\varepsilon$:

$$\Delta\varepsilon = \varepsilon(L) - \varepsilon(C). \tag{1}$$

The dielectric permittivity of leaves and needles of different types of forests depending on humidity varies for the real part of ε' from one to 40 and for the imaginary part of ε'' from 0.2 to five [6-7]. The electro physical model of tree stand as an aggregate of individual trees with a relatively low density of tree elements distribution in the volume of a stand suggests taking into account the dielectric properties of individual tree elements in a given frequency range. Anisotropy of wood determines the gradients of anisotropy as a separate tree and a forest as a whole. The random distribution of trees in space forms a heterogeneity of biomass of related groups of trees, the dimensions of which exceed by one order the characteristic dimensions of a single tree.One of the prospective areas of mathematical modeling of forest ecosystems is a tier-mosaic approach to the structuring of the forest [8]. Figure 1 illustrates the geometric structure of trees and layers of the tree stand. Within the framework of this approach, it is possible to create a dielectric model of tree stand that allows describing the patterns of space-time changes of its electro physical parameters that are sufficient to the evolution processes of the biometric characteristics of the forest.

Scaling of the basic elements of the tier-mosaic electro physical model for the tree stands in the L1- band is determined by the wavelength of the probing signal $\lambda \approx 0.2$ m.

Fig. 1. Structural levels of pine and burch tree stand: a) free bound, b) bound after cutting down.

Mature boreal forests in the conditions of the ground location of the receiving antenna allow us to distinguish the following levels of basic elements:

- level 1, needles, leaves, branches with the diameter of $d \ll \lambda$;
- level 2, branches and trunks with the diameter of $d \geq \lambda$;
- level 3, crown, tree trunk with vertical extent of $h \gg \lambda$;
- level 4, layers of crowns and tree stands with a length vertically, $H(cr)$ and $H(st) \gg \lambda$.
- level 5, groups of trees on sites up to 100 m^2

The type of a single tree, its taxonomic indices, dielectric properties of its elements determine the electro physical characteristics of the stand.

The isolated layers of the tree stand differ in terms of the volume density of the phytomass, the anisotropy of the effective permittivity, the magnitude and direction of the electro physical characteristics gradients in the vertical and horizontal directions.

A tree as an "atomic" element of the tree stand is the main structural unit of any forest and separate trees are described by structural levels 1-3. Specific differences in these structural levels are observed at any age of the trees. The tree trunk together with the large branches for a single tree represents an ordered system that can be characterized by an orientation parameter of order S [9] with respect to the vertical direction:

$$S = <3\cos 2\theta - 1>/2 \qquad (2)$$

Here θ is the angle between the branches and the vertical tree trunk. The averaging is carried out along the orientation of the whole ensemble of branches. The dispersion of the quantity S is proportional to $\cos^4\theta$. *I*t affects on cross section dispersion of this system.

The tree trunks of are oriented almost strictly vertically so the layer of trunks has a value of S that is close to one. The value of the difference in the real parts of the longitudinal and cross components of the permittivity $\Delta\varepsilon' = \varepsilon'(L) - \varepsilon'(C)$ of wood with a moisture content of 40% for spruce and larch at 20 ° C [5] are 2.8 and 9.6 respectively. Consequently, the tree

trunk layer has sufficient anisotropy to affect the right circular polarization of GPS and GLONASS signals.
Crones of trees have some tiers of branches, which differ noticeably in angle θ. The angles of the inclination of the branches for different types of trees as can be seen from Fig. 1 vary in the range from 30° to 90°. The gradients of the order parameter of the crown branches from the upper boundary of the crowns to the bottom one characterize the change in the magnitude of the anisotropy of the crowns. In the case of equally probable orientation in the space of the crown branches, the real part is $\varepsilon' = [\varepsilon'(L) + \varepsilon'(C)] / 2$, $\Delta\varepsilon' = 0$, and the crowns are a dielectric isotropic medium.
The creation of an adequate model for the distribution of navigation satellites signals in a forest canopy requires the determination of the geometric characteristics of the tree stand: the height of trees, the diameter of the trunks and the number of trees per unit area of the forest of the effective height of the forest canopy and crown. An example of forest taxation parameters of larch stand plantings near the station "Pogorel'skii Bor" of V.N. Sukachev Institute of Forest of FRC KSC SB RAS is shown in Figure 2. The experimental site of 800 m^2 contained 66 trees of different diameters and heights with crowns of different lengths. The choice of the effective thickness of the crown layer is possible with the assumption of a gradient of the crowns biomass along the vertical at the upper and lower boundaries.

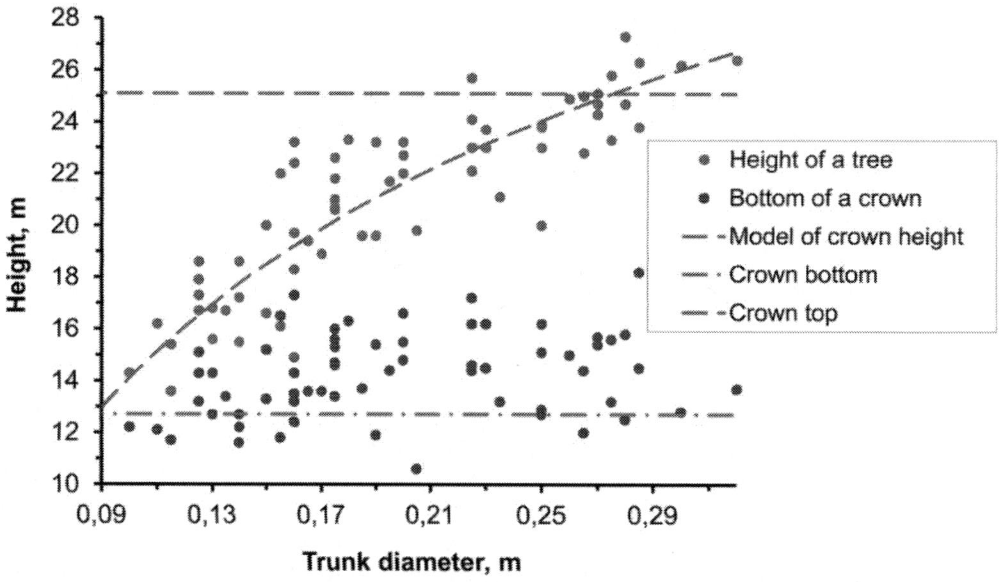

Fig. 2. Dependence of the vertical extent of tree crowns on larch on larch forest.

The boundary of the top crowns layer is formed by a trees different heights with greater depth of vertical gradient of wood distribution. The specificity of the two-layered dielectric model of the stand pertains to the mutual intersection of the layers of the trunks crowns. In fact, it is possible to introduce a third, gradient transition layer containing trunks and branches with similar in meaning diameters. The fifth-level heterogeneity retain the features of the EDP discussed above with variations, which are due to different tree densities in local groups of trees. Fig. 3 presents a satellite image and a tree distribution plan for birch and pine forests showing the presence of localized groups of trees on sites with a typical size of 10-20 m.

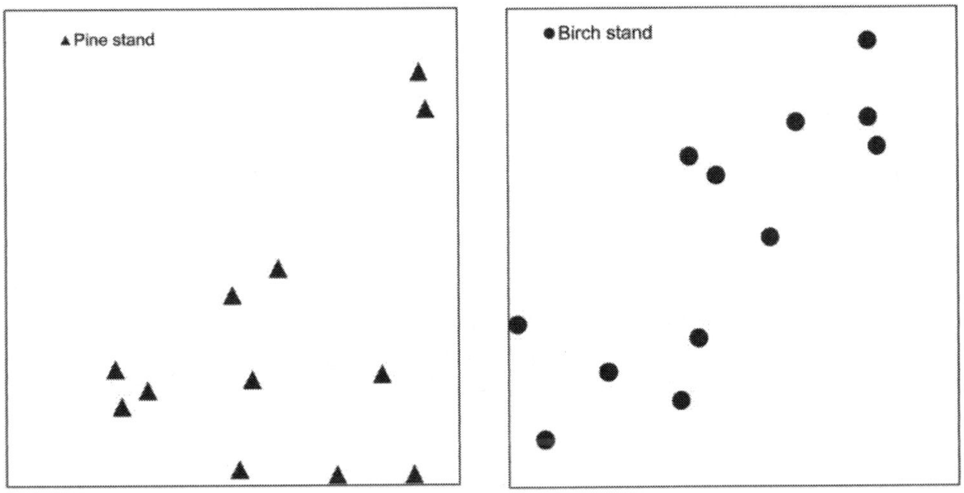

Fig. 3. Trees distribution at the experimental site.

Experimental researches of forests by radio transmission method using GNSS signals [2] revealed not monotonous dependencies of the relative amplitude on scanning trace. Scanning the forest with GNSS signals detects localized groups of trees over the anomalies dependence of the signals amplitude on time.

Figure 4 shows the recording of GLONASS 24 satellite signals when scanning pine and birch forests. Reducing the amplitude of the signal after passing through the tree stand allows us to establish the boundaries of localized groups of trees. Colors are: black (1) - correspond to signal trajectories passing in pine forest, green (2) scanning unforested-space, red (3) – birch forest.

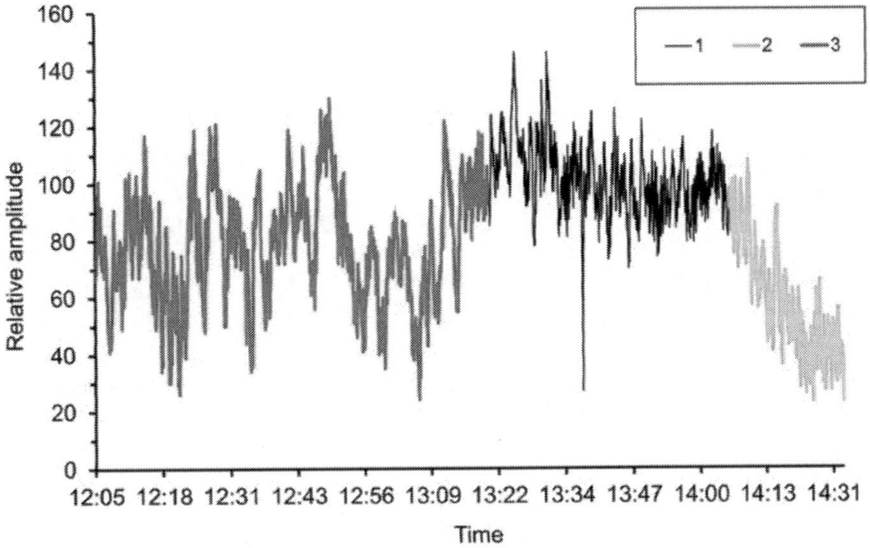

Fig.4. Fluctuations in the amplitude of GNSS signals in pine forest (1), unforested-space (2), and in birch forest (3).

3. Conclusion

A five-level structural dielectric model of the forest has been proposed taking into account the gradients of the biomass distribution and the complex effective permittivity of the tree stand.
The multi-scale structural formations of individual trees and the tree stand as a whole are distinguished which form a layer mosaic-tiered system of non-interacting components of the mixed dielectric.
The possibility of taking into account the anisotropy of the EDP in the tree crown layer with the use of the orientation order parameter of the branches is considered.
The assessments of the EDP gradients of the tree stand are made due to the inhomogeneous distribution of trees phytomass in the volume of the forest.
The possibility of using the orientational order parameter of the branches in the crown layer for estimating the gradients of the inhomogeneous EDP distribution and its spatial anisotropy is analyzied.
The prospects of using an anisotropic layered model of tree stand for describing the transformation of GPS and GLONASS signals in the forest with the purpose of restoring his biometric characteristics are discussed.

References

1. S. Jin, E. Cardellach, F. Xie, *GNSS Remote Sensing* - Springer Dordrecht Heidelberg, New York, London. - 286 p. (2014)
2. A.V. Sorokin, V.G. Podoprigora, S.V. Fomin, I.V. Savin, Evolution of the signal characteristics of navigation satellites in the forest. Izvestiya vuzov. Physics.. **60**. 12/2. pp. 19-22, (2017)
3. A.A. Chukhlantsev, A.M. Shutko, S.P. Golovachev, Radiotekhnika I Electronika. **48**, 11, pp. 1285–1311. (2003)
4. V.I. Popov, *Propagation of radio waves in forests* - M.: Hot line - Telecom,.- 392 p. (2015)
5. G.I. Torgovnikov, *Dielectric Properties of Wood and Wood Based Materials.* – Berlin, Springer – Verlag. — 194 p. (1993)
6. S.N. Novik, A.S. Zavyalov, E.D. Telpukhovsky, Electrophysical and dissipative characteristics of the properties of the elements of a coniferous stand // Izvestiya Vuzov. Physics. **49**, 7, P.57-67. (2006)
7. S.N. Novik, V.L. Mironov; V.P. Yakubov, E.D. Telpukhovsky, Height Dependence of Electromagnetic Field inside the Forest Canopy at Meter and Decimeter Wavelengths // Proc. IGARSS'06, Denver, Colorado, , **VI**, pp.2962-2964.(2006)
8. G.P. Karev, Structural models of forest ecosystems. Siberian Ecological Journal. **4** pp. 381-396. (1999)
9. De Zhen P. *Fizika zhidkih kristalov* - Moskva, Izdatelstvo "Mir", 400 p. (1977)

Application of parallel version two-dimensional fast Fourier transform algorithm, analog of the Cooley-Tukey algorithm, for digital image processing of satellite data

Mikhail Noskov [1*], *Valeriy* Tutatchikov[1], *Mikhail* Lapchik[2], *Marina* Ragulina[2], and *Tatiana* Yamskikh [1]

[1] Siberian Federal University, Institute of Space and Information Technology, Svobodny prospect 79, Krasnoyarsk, 660041, Russian Federation

[2] Omsk State Pedagogical University, Naberezhnaya Tukhachevskogo 14, Omsk, 644099, Russian Federation

Abstract. In modern systems of remote sensing two-dimensional fast Fourier transform (FFT) has been widely used for digital processing of satellite images and subsequent image filtering. This article provides a parallel version two-dimensional fast Fourier transform algorithm, analog of the Cooley-Tukey algorithm and its implementation for processing the satellite image of Krasnoyarsk and its suburban areas.

Introduction

At present, Earth remote sensing is closely related to digital image processing, as the aerospace images that are rich in detail are commonly represented in digital form of a raster type. Adjustment of contrast and brightness, spatial filtering, Fourier transform and subsequent frequency filtering are often used to improve image quality [1]. The traditionally applied algorithm for computing two-dimensional fast Fourier transform (FFT) is the sequential application of a one-dimensional FFT, first for all rows, then for all columns. The article describes a version two-dimensional fast Fourier transform algorithm, analog of the Cooley-Tukey algorithm, with the reduced number of complex operations compared to the traditionally used algorithm. A variant of the algorithm parallelization to accelerate calculations is shown in the article. The use of the algorithms to perform image processing on digital images is considered.

The filtering procedure is implemented in several stages:

- Data reading and swapping
- Parallel computation of the Cooley-Tukey algorithm
- Filtering the obtained Fourier transform
- Reverse FFT Calculation
- Image acquisition

* Corresponding author: mvnoskov@yandex.ru

Detailed description of these stages can be found in [1]. A two-dimensional analogue of the Cooley-Tukey algorithm is considered in [2], its multidimensional variant in [3] and a more detailed description of the algorithm is provided in [4]. In this paper we present a parallel two-dimensional version of the Cooley-Tukey algorithm. First, we consider two dimensional analog of the Cooley-Tukey algorithm for computing the fast Fourier transform.

The Cooley-Tukey algorithm

Given a function $f(x,y)$ of two-dimensional periodic signal with values in the complex space, where $x,y = 0, 1, \ldots, 2^s-1$, s is a positive integer. Grayscale image that is 2^s pixels in height and width can be taken as an example of such a function. The value of the function $f(x,y)$ is a complex number, which real component is equivalent to the brightness values of the corresponding pixel with coordinates (x,y) in the range 0-255, and the imaginary component of the number is zero. Then the Fourier transform $F(u,v)$ of this signal can be represented as follows:

$$F(u,v)=\sum_{u=0}^{N-1}\sum_{v=0}^{N-1} f(x,y)\cdot e^{\frac{2\pi i(ux+vy)}{N}} \tag{1}$$

where $N=2^s$, $x,y = 0,1,\ldots,N-1$.

We convert formula (1) as follows: the coordinates x and y are decomposed into even and odd parts. Then, the initial sum can be divided into four equations:

$$\begin{aligned}
F(u,v)=\sum_{u=0}^{N-1}\sum_{v=0}^{N-1}\Bigg(& f(2x1,2y1)\cdot e^{\frac{2\pi i(u\cdot 2x1+v\cdot 2y1)}{N}} + f(2x1+1,2y1)\cdot e^{\frac{2\pi i(u\cdot(2x1+1)+v\cdot 2y1)}{N}} + \\
& + f(2x1,2y1+1)\cdot e^{\frac{2\pi i(u\cdot 2x1+v\cdot(2y1+1))}{N}} + f(2x1+1,2y1+1)\cdot e^{\frac{2\pi i(u\cdot(2x1+1)+v\cdot(2y1+1))}{N}}\Bigg) = \\
&= \sum_{u=0}^{N-1}\sum_{v=0}^{N-1} f(2x1,2y1)\cdot e^{\frac{2\pi i(u\cdot 2x1+v\cdot 2y1)}{N}} + e^{\frac{2\pi iu}{N}}\cdot\sum_{u=0}^{N-1}\sum_{v=0}^{N-1} f(2x1+1,2y1)\cdot e^{\frac{2\pi i(u\cdot 2x1+v\cdot 2y1)}{N}} + \\
&+ e^{\frac{2\pi iv}{N}}\cdot\sum_{u=0}^{N-1}\sum_{v=0}^{N-1} f(2x1,2y1+1)\cdot e^{\frac{2\pi i(u\cdot 2x1+v\cdot 2y1)}{N}} + \\
&+ e^{\frac{2\pi iu}{N}}\cdot e^{\frac{2\pi iv}{N}}\cdot\sum_{u=0}^{N-1}\sum_{v=0}^{N-1} f(2x1+1,2y1+1)\cdot e^{\frac{2\pi i(u\cdot 2x1+v\cdot 2y1)}{N}} = \\
&= g_{00}(u,v)+e^{\frac{2\pi iu}{N}}\cdot g_{10}(u,v)+e^{\frac{2\pi iv}{N}}\cdot g_{01}(u,v)+e^{\frac{2\pi iu}{N}}\cdot e^{\frac{2\pi iv}{N}}\cdot g_{11}(u,v)
\end{aligned} \tag{2}$$

Each of the sums $g(u,v)$ obtained in (2) is a FFT signal with the elements $2^{s-1}\times 2^{s-1}$ for even or odd coordinates of the source signal $f(x,y)$

Then, in the next step , we divide the coordinates u and v into two equal-sized subsets $2u1$, $2v1$ and $2u1+N/2$, $2v1+N/2$, respectively, $u1,v1 = 0,1,\ldots,N/2-1$. Then for the factors $e^{\frac{2\pi iu}{N}}$ and $e^{\frac{2\pi iv}{N}}$ from the second subset we get:

$$\begin{aligned}
e^{\frac{2\pi i\cdot(2u1+N/2)}{N}} &= e^{\frac{2\pi i\cdot(2u1)}{N}}\cdot e^{\frac{2\pi i\cdot(N/2)}{N}} = e^{\frac{2\pi i\cdot(2u1)}{N}}\cdot e^{\pi i} = -e^{\frac{2\pi i\cdot(2u1)}{N}}, \\
e^{\frac{2\pi i\cdot(2v1+N/2)}{N}} &= e^{\frac{2\pi i\cdot(2v1)}{N}}\cdot e^{\frac{2\pi i\cdot(N/2)}{N}} = e^{\frac{2\pi i\cdot(2v1)}{N}}\cdot e^{\pi i} = -e^{\frac{2\pi i\cdot(2v1)}{N}}.
\end{aligned} \tag{3}$$

 https://doi.org/10.1051/e3sconf/20197501012

We plug (3) into the formula (2) to get:

$$
\begin{aligned}
&F(u1,v1) = g_{00}(u1,v1) + e^{\frac{2\pi i\cdot u1}{N}} \cdot g_{10}(u1,v1) + e^{\frac{2\pi i\cdot v1}{N}} \cdot g_{01}(u1,v1) + \\
&+ e^{\frac{2\pi i\cdot u1}{N}} \cdot e^{\frac{2\pi i\cdot v1}{N}} \cdot g_{11}(u1,v1), \\
&F(u1+N/2,v1) = g_{00}(u1+N/2,v1) - e^{\frac{2\pi i\cdot u1}{N}} \cdot g_{10}(u1+N/2,v1) + \\
&+ e^{\frac{2\pi i\cdot v1}{N}} \cdot g_{01}(u1+N/2,v1) - e^{\frac{2\pi i\cdot u1}{N}} \cdot e^{\frac{2\pi i\cdot v1}{N}} \cdot g_{11}(u1+N/2,v1), \\
&F(u1,v1+N/2) = g_{00}(u1,v1+N/2) + e^{\frac{2\pi i\cdot u1}{N}} \cdot g_{10}(u1,v1+N/2) - \qquad (4) \\
&- e^{\frac{2\pi i\cdot v1}{N}} \cdot g_{01}(u1,v1+N/2) - e^{\frac{2\pi i\cdot u1}{N}} \cdot e^{\frac{2\pi i\cdot v1}{N}} \cdot g_{11}(u1,v1+N/2), \\
&F(u1+N/2,v1+N/2) = g_{00}(u1+N/2,v1+N/2) - \\
&- e^{\frac{2\pi i\cdot u1}{N}} \cdot g_{10}(u1+N/2,v1+N/2) - e^{\frac{2\pi i\cdot v1}{N}} \cdot g_{01}(u1+N/2,v1+N/2) + \\
&+ e^{\frac{2\pi i\cdot u1}{N}} \cdot e^{\frac{2\pi i\cdot v1}{N}} \cdot g_{11}(u1+N/2,v1+N/2),
\end{aligned}
$$

where $u1,v1 = 0,1,\ldots,N/2-1$.

Formula (4) describes the two-dimensional fast Fourier transform batterfly in analog of Cooley-Tukey algorithm. It is schematically depicted in Figure 1. With this butterfly we can split the source signal and the Fourier transform with elements into four sub-signals each with the number of elements $2^{s-1} \times 2^{s-1}$ [4]. This reduces the number of multiplications and additions of complex numbers required to compute FFT.

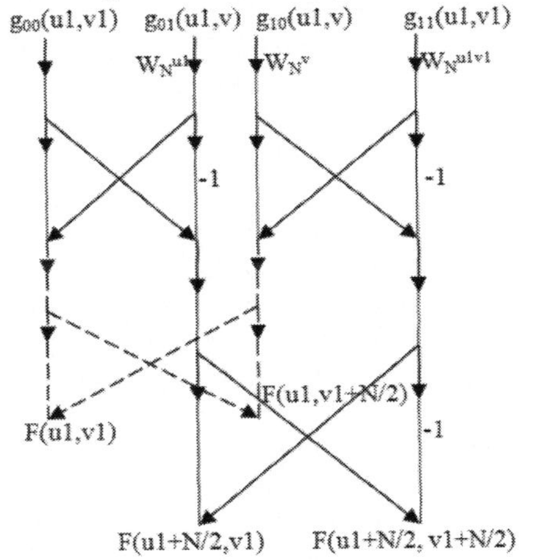

Fig. 1. Two-dimensional fast Fourier transform batterfly in analog of Cooley-Tukey algorithm

Decomposition (4) can be applied to each sub-signal *g(u,v)* several times until we get sub-signals of 4 elements, which FFTs are being computed directly. Then for the signal *f(x,y)* of the elements $N \times N$ the total number of multiplications required to compute complex numbers will be $\frac{3}{4} N^2 \log_2 N$, and the number of additions - $2N^2 \log_2 N$ [3]. For comparison, the standard method of computing two-dimensional FFT by rows and columns will require $N^2 \log_2 N$ multiplications and $2N^2 \log_2 N$ additions.

The parallel Cooley-Tukey algorithm

To test the running time of an algorithm for calculating two-dimensional fast Fourier transform algorithm, analog of the Cooley-Tukey algorithm, a simulation program was written in the C++ programming language [4]. Two principal methods of algorithm parallelization were used: OpenMP, targeted toward use on shared memory systems, and MPI for distributed memory systems. Testing was carried out on the cluster node of the SFU supercomputer with IBM HS21 XM Xeon Quad core E5450 3.0 GHz, 64 Gb RAM [5]. The result of testing for a system with distributed memory on a single cluster node is presented in Table 1.

The two-dimensional analogue of the Cooley-Tukey algorithm processes the signal with the number of samples $2^s \times 2^s$ in s iterations. In the first iteration, due to the preliminary permutation of the elements, the signal data is divided into quads of related (vertically and horizontally) elements spaced by one element from each other; in the second iteration, into quads of elements separated by two elements from each other; in the third iteration - by the 2^2 element; in the last s- iteration by- 2^{s-1} element. For such implementation of the algorithm, in shared memory systems parallel data structures are designed by splitting multiple data in each iteration into arrays of interconnected sets of the elements for each separate thread. In a distributed memory systems, a similar partitioning into sets of related data occurs at each iteration with subsequent transfer of data between the processes for independent computations.

Table 1. The result of testing a two-dimensional parallel analogue of the Cooley-Tukey algorithm in a system with distributed memory, in seconds

Size	The number of processes	2D FFT, for rows and columns	2D Cooley-Tukey FFT
1024*1024	1	0,490	0,310
	2	0,310	0,310
	4	0,220	0,270
	8	0,180	0,270
	16	0,180	0,350
2048*2048	1	2,300	1,330
	2	1,550	1,260
	4	1,060	0,950
	8	0,840	0,850
	16	0,850	1,000
4096*4096	1	9,880	5,850
	2	6,240	4,590
	4	4,370	3,590
	8	3,430	3,060
	16	3,740	3,390
8192*8192	1	43,210	25,190
	2	26,990	19,160
	4	18,550	14,870
	8	14,520	13,130
	16	14,090	11,700

The data are presented graphically in Figure 2, where the FFT algorithm by rows and columns is denoted by FFT RC, and the analogue of the Cooley-Tukey algorithm by FFT CT.

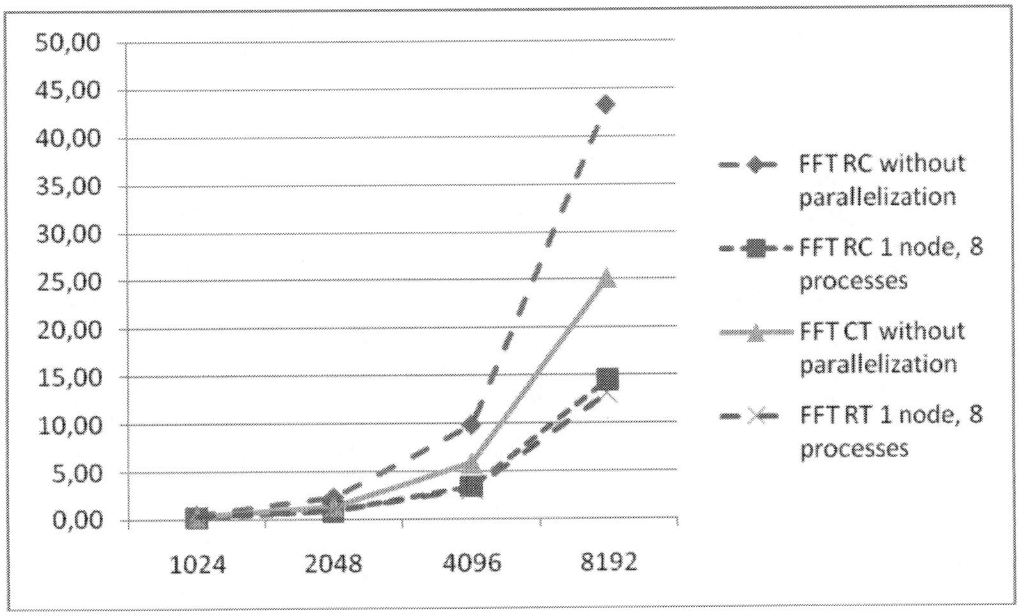

Fig. 2. Comparison of running time for various algorithms

Image filtering

LandSat-8 image of Krasnoyarsk and its surrounding areas dated April 7, 2016 [6], which is shown in Figure 1 on the left, was used as a test signal. The original image resolution is 8081 * 8171 pixels, it was converted to the nearest power of two: and then scaled for the powers 10-15. On the right side of the figure 3 the result from high-pass filtering is shown. In this case, the contours are more vivid: the river line, the boundaries of the rocky areas.

Fig. 3. Original image - The result from high-pass filtering

The original image is shown on the left side of the figure 4. The result from low-pass filtering is on the right side. In this case, small sharp changes in mountainous terrain are not so noticeable on the general background, that is, small details have been removed.

Fig. 4. Original Image - Result from low pass filtering

Conclusion

It is shown that the parallel analogue of the Cooley-Tukey algorithm for computing two-dimensional FFT is executed on average 4 times faster than the standard method for computing two-dimensional fast Fourier transform by rows and columns.

References

1. R.C. Gonzalez and R.E. Woods, *Digital Image Processing*, Second Edition, Pearson Education, Inc (2004)
2. R. Blahut, *Fast algorithms for digital signal processing*, Addison-Wesley Publishing Company (1985)
3. V.S. Tutatchikov, O.I. Kiselev, M.V. Noskov, *Calculating the n-Dimensional Fast Fourier Transform,* PRIA, **23(3)**, 429-433 (2013)
4. V.S. Tutatchikov, *Two-dimensional fast Fourier transform, Proceeding of 11th International Forum on Strategic Technology (IFOST-2016)*, 495 – 498 (2016)
5. Supercomputer complex of SFU Access mode: http://cluster.sfu-kras.ru/
6. LandSat-8 image of Krasnoyarsk and its surrounding areas Access mode: http://earthexplorer.usgs.gov/

Satellite imagery of lengthy territories with complex configuration with the account of attitude and positioning errors

Dmitriy Mozgovoy[1,*], *Dmitriy* Svinarenko[1], *Roman* Tsarev[2], and *Tatiana* Yamskikh[2]

[1]Oles Honchar Dnipropetrovsk National University, 72, Gagarin Prospect, Dnipropetrovsk, 49000, Ukraine
[2]Siberian Federal University, 79, Svobodny Prospect, Krasnoyarsk, 660041, Russia

Abstract. A method for monitoring attitude and positioning errors when taking satellite imagery of lengthy territories with complex configuration using an ultra-high spatial resolution optical-electronic scanner is described in the article. The results of modeling the system of automatic satellite attitude program control during the process of imagery are presented. Given these results, the impact of attitude and positioning errors during satellite imagery was estimated on the coverage percentage of the territory to be imaged.

1 Introduction

Satellite imagery is the most effective way to obtain spatial data due to its visibility, reliability, efficiency, contactless detailed observation and other advantages compared with traditional ground-based methods [1-6]. The new method of imagery in the process of re-targeting (i.e., with non-zero angular velocities), implemented using satellite attitude program control, can significantly increase the efficiency and accuracy of satellite imagery of lengthy territories with complex configuration. This method of satellite imagery provides possibilities for:

- performing imagery of arbitrarily-spaced lengthy territories with complex configuration in one-orbit period (i.e., to increase the efficiency of imaging a given territory);
- increasing the amount of relevant remote sensing data obtained in one session (i.e., to increase the percentage coverage of a given territory);
- taking images from neighboring orbits at low angles of the Sun at a given signal-to-noise ratio;
- detecting low-contrast objects in the images under given lighting conditions;
- increasing the signal-to-noise ratio for satellite images and improve image classification procedures using radiometric and spectral characteristics.

Moreover, when taking satellite imagery in the retargeting mode, the duration of communication session in the direct transmission mode (for the same territory being

[*] Corresponding author: m-d-k@i.ua

imaged) increases, thus making it possible to reduce the data rate and use receiving stations with antennas having lower gain and smaller size (with the same parameters of onboard transmitting equipment).

2 Limitations of the proposed method

The use of satellite attitude program control of a satellite angular motion imposes increased requirements on the on-board attitude system both in terms of providing the required attitude accuracy and stabilization of the satellite, and in providing the required retargeting velocity directly during the imagery. Moreover, special requirements apply to the subsystem of magnetic discharge of inertial energy-storage devices (flywheels). Therefore, there arose a necessity for computer modeling of new satellite imaging modes with ultra-high spatial resolution scanners, given the attitude and positioning errors.

3 The main stages of research

The main stages of modeling satellite imagery of lengthy territories with complex configuration are as follows:

- approximation of a lengthy object defined by separate nodal points on a digital map (linear, quadratic, spline, etc.);
- determination of the optimal coverage of a lengthy object, taking into account the swath width of the imaging instrument and the required scanning direction;
- calculation of the orbital motion in order to choose the orbit for imagery and the time for switching on the imaging instrument, taking into account the Delay coefficient (when implementing the "Time Delay and Integration" mode);
- calculation of the satellite attitude angles required for imaging a given lengthy object, taking into account the installation angles of the imaging instrument;
- estimation of the satellite angular velocities during the imagery and analysis of the imagery feasibility, taking into account the existing limitations on the values of pointing angles and satellite angular velocities;
- modeling of satellite imagery process, taking into account additional factors (atmospheric refraction, terrain relief displacement, etc.);
- estimation of the impact of systematic and random attitude and positioning errors on agility and coverage percentage.

4 The description of mathematical models used

4.1 Mathematical model of a lengthy territory with complex configuration

The territory with complex configuration to be imaged is marked on the map by nodal points with arbitrary step in latitude and longitude. To approximate the lengthy territory, a natural cubic interpolating spline is used, then smoothed with the least squares method.

4.2 Determination of the required satellite attitude in the process of imagery

The input data are composed of:

- coordinates of the reference points in geographic coordinate system corresponding to start and end points of the line;
- satellite coordinates in geographic coordinate system;

- pointing vectors of the reference points in the instrument coordinate system;
- the values of the elements in the transition matrix from the orbital coordinate system to the geographic coordinate system.

Yaw, pitch and roll attitude angles and angular velocities identify a reference frame that moves with the satellite.

The calculation procedure is as follows:

- normalised vectors of the reference points alignment in geographic coordinate system are calculated;
- the values of the elements in the transition matrix from the geographic coordinate system to the orbital coordinate system are determined;
- vectors of the reference points alignment in the orbital coordinate system are calculated;
- the values of the elements in the transition matrix from the instrument coordinate system to the orbital coordinate system are defined;
- yaw, pitch and roll attitude angles are determined by the values of the elements in the transition matrix from the instrument coordinate system to the orbital coordinate system.

4.3 The model of satellite orbital motion

For short-term prediction and low earth circular orbits, sufficient accuracy is achieved with the propagator SGP4, which allows to work with the initial conditions in TLE files generated by NORAD.

4.4 Review and analysis of the impact of satellite attitude and positioning errors

The impact of random errors that occur when setting attitude parameters and satellite tracking on the feasibility and efficacy of the imagery program in terms of the coverage accuracy was estimated using the finite differences method.

The input data are composed of the errors in specifying the attitude angles and prediction of the satellite position. The output data are the displacements of the reference points' cartographic coordinates corresponding to the maximum attitude errors and the position of the satellite.

The peculiarity of the method consists in calculating the values of partial differentials from the reference points' cartographic coordinates in the satellite coordinates and the roll yaw and pitch angles. They are calculated for all combinations of maximum and minimum absolute values of errors in specifying satellite attitude angles and prediction of the satellite position corresponding to start and end points of all lines in the scene being imaged.

5 The input data for modeling

The input data for modeling satellite imagery of lengthy territories with complex configuration are as follows:

- the territory to be imaged is the southern coast of Crimea from Sevastopol to Feodosiya;
- the type of satellite orbit - solar-synchronous,
- the altitude of the satellite orbit is 700 km;
- the scanner swath width at nadir - 20 km.

As a basis for modeling we used the type of imagery without TDI (Time Delay and Integration) implementation when satellite attitude errors were not considered in simulation, as well as with the maximum satellite attitude errors ± 0.1 ° in roll, pitch and yaw.

6 The results of modeling

The results of modeling satellite imagery of lengthy territories with complex configuration (the scanner swath width - 20 km) when satellite attitude errors were not considered are shown in Figure 1. Figure 2 presents the results of modelling with the maximum satellite attitude errors.

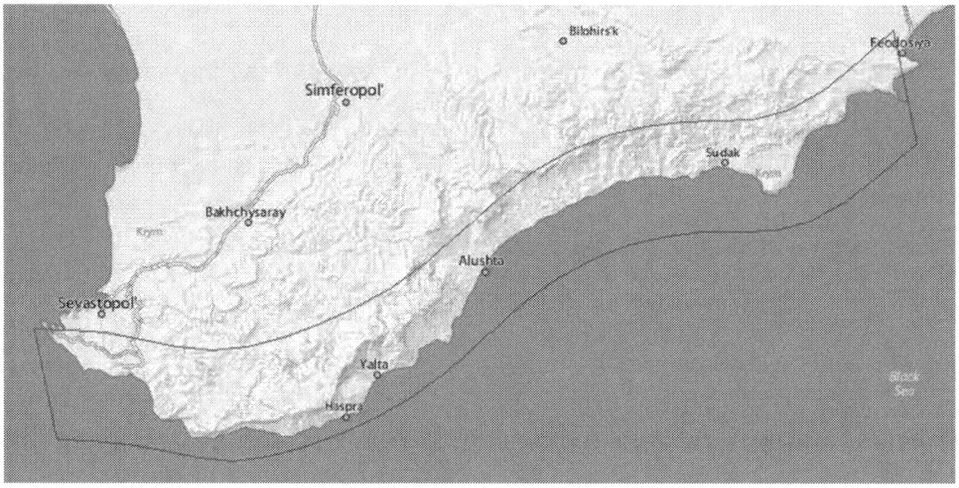

Fig. 1. Modeling satellite imagery when satellite attitude errors are not taken into account.

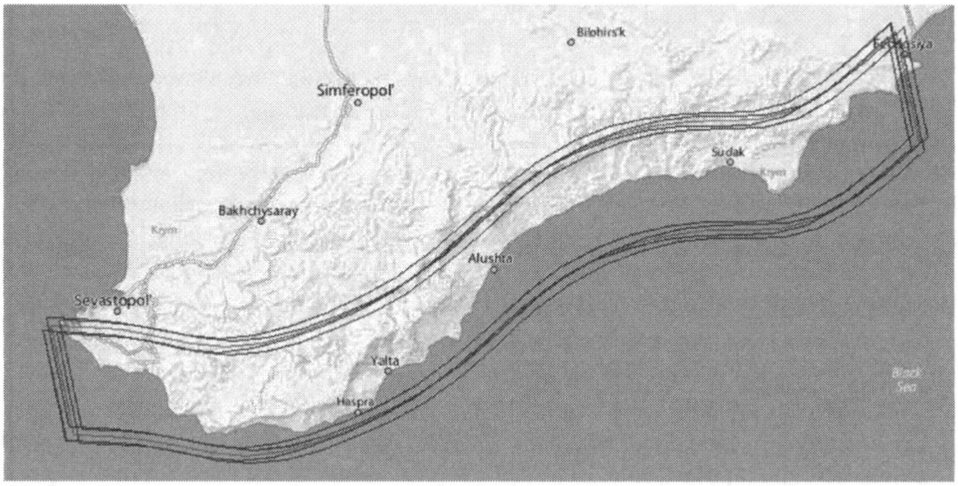

Fig. 2. Modeling satellite imagery with the account of the maximum satellite attitude errors ± 0.1 ° in roll, pitch and yaw.

The simulation results of satellite imagery of the lengthy coastline on the southern coast of Crimea from Sevastopol to Feodosiya with measurement data of cloud amount for a given area showed, that in a given 5-day period, the territory to be imaged was completely available for 1 day, partially available for 2 days, unavailable for 2 days.

The results of simulating the impact of satellite attitude and positioning errors when scanning a lengthy coastline showed that with a maximum satellite attitude error of ± 0.1 ° for roll, pitch and yaw, about 90% and 80% of the territory was covered with the scanner swath width 40 and 20 km respectively. The satellite orbit prediction errors did not reduce the percentage of guaranteed coverage due to their insignificant values (sub-pixel mapping was used for a given scale).

7 Evaluation of the method for satellite imagery of lengthy territories with complex configuration

Table 1 represents the parameters determining the efficiency of modeling for various types of satellite imagery of the lengthy coastline on the southern coast of Crimea from Sevastopol to Feodosiya taking into account measurement data of cloud amount for a given area and satellite attitude and positioning errors.

Table 1. The basic parameters determining the efficiency of modeling for various types of satellite imagery taking into account measurement data of cloud amount for a given area and satellite attitude and positioning errors

Type	The number of scenes (orbits)	Task duration, days	Coverage
Imagery with scene orientation along the subsatellite track direction	3	5	52%
Imagery with an arbitrary scene orientation	1	1	80%
Imagery of scenes with complex configuration	1	1	98%

In the process of satellite imagery of lengthy territories with complex configuration, the new method gave an increase in the effective area of about 15%. Weather conditions are more important for autumn-winter period. Thus, the comparative efficiency of the method developed for satellite imagery will be even higher. Improving the efficiency of imagery by means of satellite attitude program control during imagery provides possibilities for:

- reducing the time required for taking images of arbitrarily-spaced lengthy territories with complex configuration;
- increasing the signal-to-noise ratio of satellite images when taking images from neighboring orbits at low angles of the Sun;
- increasing the amount of relevant remote sensing data obtained in one session.

8 The technique for estimating requirements to attitude and stabilization subsystems of remote sensing satellites

On the basis of the developed mathematical models, we developed a technique for estimating requirements to attitude and stabilization subsystems of remote sensing satellites with optical-electronic scanners of ultrahigh spatial resolution, which increases the efficiency and reliability of data obtained while scanning small-sized objects located on lengthy territories with complex configuration. Based on the requirements to the agility and the percentage coverage of the lengthy territory to be imaged, this technique provides possibilities for the development of requirements to attitude and stabilization subsystems of remote sensing satellites with optical-electronic scanners of ultrahigh spatial resolution at the design stage.

9 Conclusion

The results of modeling satellite imagery of lengthy territories with complex configuration, with the account of satellite attitude and positioning errors, confirmed the efficiency of automatic satellite attitude program control during the process of imagery.

Simulated satellite imagery of the lengthy coastline on the southern coast of Crimea from Sevastopol to Feodosiya with the account of measurement data of cloud amount for a given area and satellite attitude and positioning errors, demonstrated high agility of the method proposed (1 day) and higher percentage of the effective area (98%) compared to traditional satellite imagery (5 days and 52% respectively).

The series of simulation results allowed us to conclude that one-orbit imagery of a scene characterized by complex configuration using satellite attitude program control during the process of imagery is the most effective type of satellite imagery in terms of agility and coverage percentage.

References

1. P.C. Gray, J.T. Ridge, S.K. Poulin, A.C. Seymour, A.M. Schwantes, J.J. Swenson, D.W. Johnston, Remote Sens., **10,** 1257 (2018)
2. X. Huang, D. Wen, J. Li, R. Qin, Remote Sens. Environ., **196**, 56-75 (2017)
3. Y. Kim, S.B. Lee, H. Yun, J. Kim, Y. Park, Int. Geoscience and Remote Sensing Symposium (IGARSS), 4862-4865 (2017)
4. J. Liu, W.J. Emery, X. Wu, M. Li, C. Li, L. Zhang, IEEE T. Geosci. Remote, **55**, 7113-7125 (2017)
5. M.S. Moussavi, W. Abdalati, A. Pope, T. Scambos, M. Tedesco, M. MacFerrin, S. Grigsby, Remote Sens. Environ., **183**, 294-303 (2016)
6. O.L. Makarov, D.K. Mozgovoy, V.S. Khoroshilov, 21th Int. Conf. on Microwave and Telecommunication Technology (CriMiCo), 905-907 (2011)

Simulated satellite imagery of lengthy territories with complex configuration

Dmitriy Mozgovoy[1,*], *Dmitriy* Svinarenko[1], *Roman* Tsarev[2], and *Tatiana* Yamskikh[2]

[1] Oles Honchar Dnipropetrovsk National University, 72, Gagarin Prospect, Dnipropetrovsk, 49000, Ukraine
[2] Siberian Federal University, 79, Svobodny Prospect, Krasnoyarsk, 660041, Russia

Abstract. The study presents the results of creating simulated satellite imagery of lengthy territories with complex configuration using data from satellites with optical-electronic scanners of ultrahigh spatial resolution. The article attempts to clarify the feasibility of one-orbit imagery of a given lengthy object, taking into account the existing limitations on the values of pointing angles and satellite angular velocities based on the results of modeling the system of automatic satellite attitude program control during the process of imagery.

1 Problem statement

When solving a number of economic and scientific problems, there arose a need for fast satellite imagery of lengthy territories with complex configuration (highways and railroads, oil and gas pipelines, border areas, navigable waterways, coastlines, etc.) [1-6]. For this purpose ultrahigh resolution remote sensing satellites are commonly used. The distinctive features of Earth remote sensing satellites with optical-electronic scanners providing ultrahigh spatial resolution are as follows [1, 7-10]:

1) a narrow swath width, which does not allow to take images of arbitrarily-spaced lengthy territories in one-orbit period;

2) a small pixel size of CCD photocells (i.e., low sensitivity), which does not allow to make images of low-contrast objects, as well as take pictures from neighboring orbits at low angles of the Sun.

To cover a lengthy object with an arbitrary configuration, it is necessary to take several images from different orbits, which takes from 3 days to 3 weeks. It may take even longer time due to cloudiness, which is unacceptable for most tasks.

2 Possible solutions

In the remote sensing satellites with optical-electronic scanners of ultra-high spatial resolution existing abroad, the following problems have been partially solved:

* Corresponding author: m-d-k@i.ua

- some satellites allow to take images of arbitrarily oriented straight-line lengthy territories (Fig. 1, in the middle), choosing the most optimal scanning direction (however, imagery of a lengthy territory with complex configuration is still performed in several orbits);

- for some satellites it is possible to take images increasing the exposure time (i.e., take pictures of low contrast or low light objects), but in this case the scene has a shape of a rectangle.

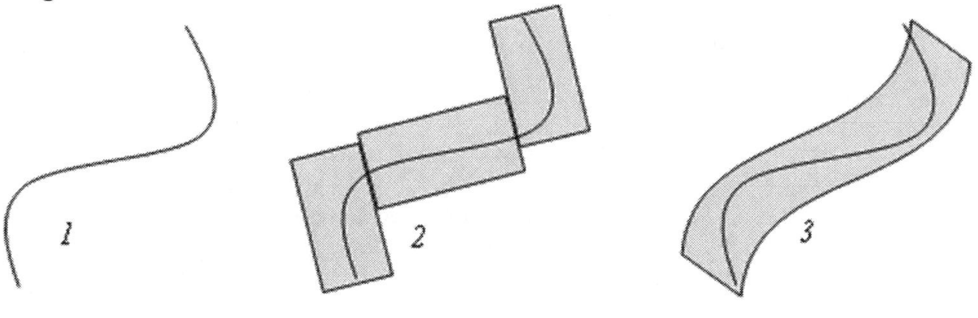

Fig. 1. Options for satellite imagery of the lengthy object
(1 – Target; 2 – Three-orbit imagery; 3 – One-orbit imagery)

3 The nature of a new method of satellite imagery

An advanced technique implemented using automatic satellite attitude program control in the process of retargeting (i.e., with non-zero angular velocities) may significantly increase the efficiency of satellite imagery of lengthy territories with complex configuration.

In this case, it is possible to choose the best scan mode to take images of arbitrarily-spaced lengthy territories in one-orbit period, i.e. much faster (Fig. 1, on the right). Moreover, non-zero angular velocities allow to increase the exposure time and thus improve image classification procedures using radiometric and spectral characteristics.

4 Limitations of the proposed method

The use of automatic satellite attitude program control system of satellite angular motion imposes increased requirements on the on-board attitude system both in terms of providing the required attitude accuracy and stabilization of the satellite, and in providing the required retargeting velocity directly during the imagery. Moreover, special requirements apply to the subsystem of magnetic discharge of inertial energy-storage devices (flywheels).

Given these limitations, caused by a more intensive retargeting of the satellite directly in the process of imagery, and, as a result, increased power consumption, there arose a necessity for computer modeling of new satellite imaging modes with ultra-high spatial resolution scanners.

5 The main stages of research

The main stages of modeling satellite imagery of lengthy territories with complex configuration:

- approximation of a lengthy object defined by separate nodal points on a digital map (linear, quadratic, spline, etc.);

- determination of the optimal coverage of a lengthy object, taking into account the swath width of the imaging instrument and the required scanning direction;
- calculation of the orbital motion in order to choose the orbit for imagery and the time for switching on the imaging instrument, taking into account the Delay coefficient (when implementing the "Time Delay and Integration" mode);
- calculation of the satellite attitude angles required for imaging a given lengthy object, taking into account the installation angles of the imaging instrument;
- estimation of the satellite angular velocities during the imagery and analysis of the imagery feasibility, taking into account the existing limitations on the values of pointing angles and satellite angular velocities;
- modeling of satellite imagery process, taking into account additional factors (atmospheric refraction, terrain relief displacement, etc.);
- estimation of the impact of systematic and random errors on the accuracy of obtaining the object coordinates from the image (errors in the installation of the imaging instrument, errors in determining the satellite attitude and position parameters, etc.).

6 The description of mathematical models used

6.1 Mathematical model of a lengthy territory with complex configuration

The territory with complex configuration to be imaged is marked on the map by nodal points with arbitrary step in latitude and longitude. To approximate the lengthy territory, a natural cubic interpolating spline is used, then smoothed with the least squares method.

6.2 Determination of satellite attitude in the process of imagery

The input data are composed of:
- coordinates of the reference points in geographic coordinate system corresponding to start and end points of the line;
- satellite coordinates in geographic coordinate system;
- pointing vectors of the reference points in the instrument coordinate system;
- the values of the elements in the transition matrix from the orbital coordinate system to the geographic coordinate system.

Yaw, pitch and roll attitude angles and angular velocities identify a reference frame that moves with the satellite.

The calculation procedure is as follows:
- normalised vectors of the reference points alignment in geographic coordinate system are calculated;
- the values of the elements in the transition matrix from the geographic coordinate system to the orbital coordinate system are determined;
- vectors of the reference points alignment in the orbital coordinate system are calculated;
- the values of the elements in the transition matrix from the instrument coordinate system to the orbital coordinate system are defined;
- yaw, pitch and roll attitude angles are determined by the values of the elements in the transition matrix from the instrument coordinate system to the orbital coordinate system.

Angular velocities are found by analytical differentiating the coefficients of the spline interpolating the attitude angles.

6.3 The model of satellite orbital motion

For short-term prediction and low earth circular orbits, sufficient accuracy is achieved with the propagator SGP4, which allows to work with the initial conditions in TLE files generated by NORAD.

7 The input data for modeling

The input data for modeling satellite imagery of lengthy territories with complex configuration are as follows:

- the territory to be imaged is 600 km in length;
- the type of satellite orbit - solar-synchronous, with about 700 km height;
- the scanner swath width at nadir - 20 km.

The following types of imagery were modeled:

- without implementation of TDI (Time Delay and Integration) mode (spline approximation of the survey object);
- using TDI mode (CTDI = 4, spline approximation of the survey object);
- without implementation of TDI mode (linear approximation of the survey object);
- using TDI mode (CTDI = 4, linear approximation of the survey object).

8 The results of modeling

The results of satellite imagery modeling of lengthy territories with complex configuration using TDI mode are shown in figures 2...4 (angles in degrees, angular velocity in degrees/sec, time in sec).

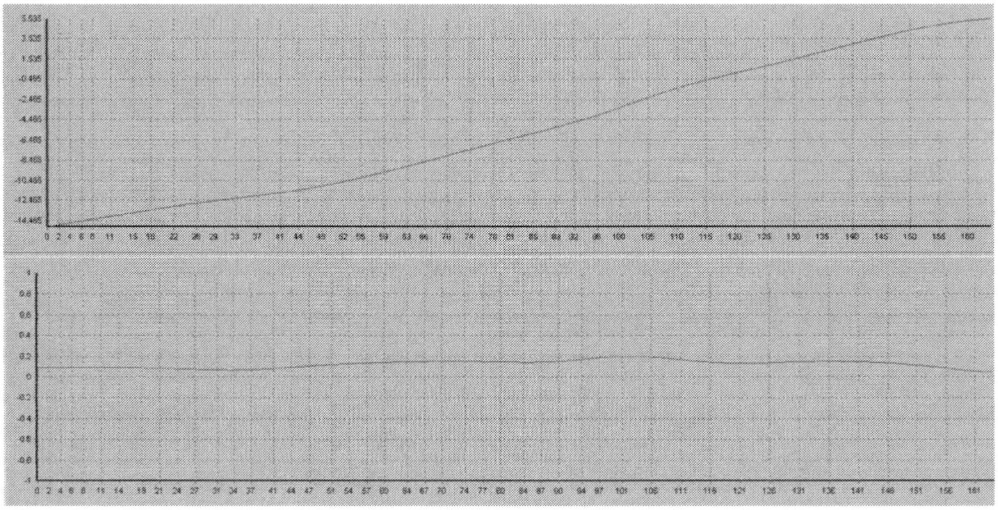

Fig. 2. Roll attitude angle (upper figure) and roll angular velocity (lower figure) during imagery using TDI mode (spline approximation of the survey object).

The results of modeling have shown that in one-orbit period it is possible to provide full coverage of the specified territory with spline approximation of the survey object, while linear approximation of the object being imaged allows to scan the required territory only partially, although at lower angular speeds.

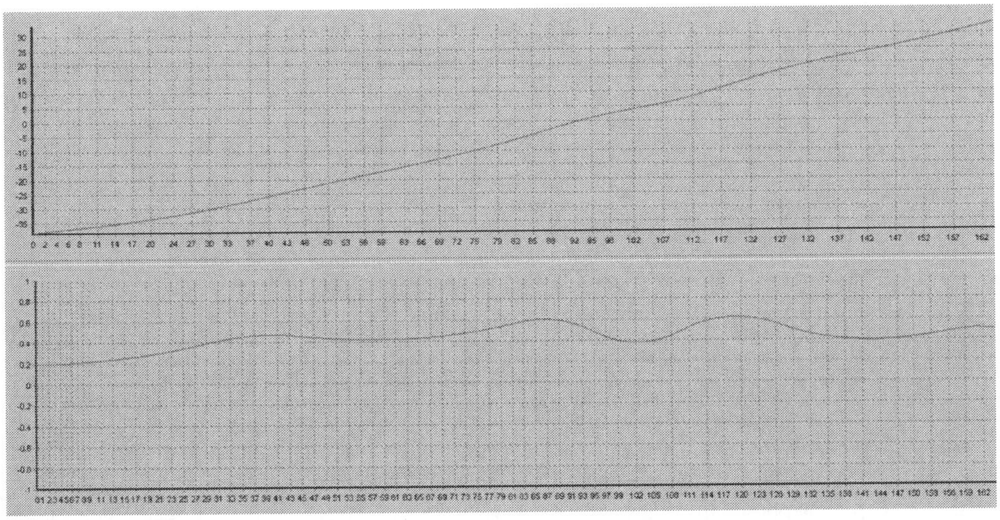

Fig. 3. Pitch angle (upper figure) and pitch angular velocity (lower figure) during imagery using TDI mode (spline approximation of the survey object).

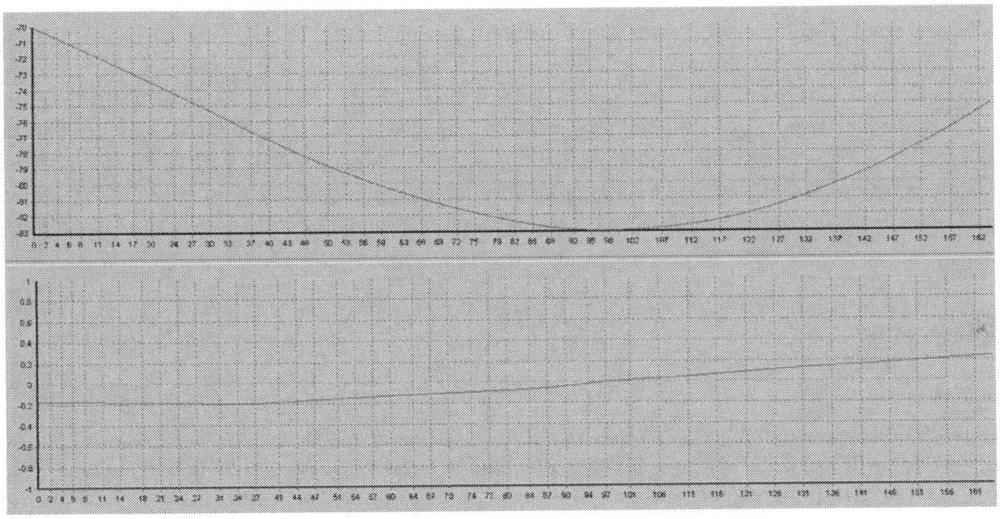

Fig. 4. Yaw angle (upper figure) and yaw angular velocity (lower figure) during imagery using TDI mode (spline approximation of the survey object).

Table 1 shows the ranges of attitude angles and satellite angular velocities required for various types of imagery, obtained as a result of modeling.

Table 1. The ranges of attitude angles and satellite angular velocities

Approxi-mation type	Parameters of attitude and satellite angular motion	Unit of measu-rement	Imagery without implementation of TDI mode (the duration of imagery 46 sec)		Imagery using TDI mode (the duration of imagery 164 sec)	
			Min value	Max value	Min value	Max value
Spline (coverage 100%)	Roll angle	degree	-16.6	5.4	-14.8	5.5
	Pitch angle	degree	-11.9	14.2	-40	34
	Yaw angle	degree	-81	-79.1	-83	-73
	Roll velocity	degree/sec	0.29	0.65	0.08	0.21
	Pitch velocity	degree/sec	-0.05	1.4	0.2	0.62
	Yaw velocity	degree/sec	-0.21	0.39	-0.2	0.23
Linear (coverage 85%)	Roll angle	degree	-17.9	9	-14.7	5.9
	Pitch angle	degree	-13.1	12.8	-39	33
	Yaw angle	degree	-83.7	-80.9	-83.1	-70.4
	Roll velocity	degree/sec	0.45	0.52	0.05	0.18
	Pitch velocity	degree/sec	0.51	0.58	0.29	0.52
	Yaw velocity	degree/sec	-0.22	0.09	-0.2	0.21

9 Conclusion

Computer modeling has confirmed the possibility of satellite imagery of lengthy territories with complex configuration at non-zero angular velocities using automatic satellite attitude program control system in the process of imagery. The angular velocities of satellite retargeting averaged less than 0.5 deg/sec. The spline approximation of the survey object appeared to be the most effective type of approximation in terms of agility (in one-orbit period) and coverage percentage.

References

1. Z. Chen, J. Pasher, J. Duffe, A. Behnamian, Can. J. Remote Sens., **43**, 513-527 (2017)
2. Y.-S. Hsiao, C. Hwang, Y.-S. Cheng, L.-C. Chen, H.-J. Hsu, J.-H. Tsai, C.-L. Liu, C.-C. Wang, Y.-C. Liu, Y.-C. Kao, Remote Sens. Environ., **176**, 69-83 (2016)
3. P.-C. Hsu, M.-H. Chang, C.-C. Lin, S.-J. Huang, C.-R. Ho, Remote Sens. Environ., **193**, 54-64 (2017)
4. D.R. Rounce, C.S. Watson, D.C. McKinney, Remote Sens., **9**, 654 (2017)
5. A. Shaker, W.Y. Yan, S. Easa, Geocarto. Int., **26**, 49-67 (2011)
6. A.M. Youssef, B. Pradhan, N.H. Maerz, Arab. J. Geosci., **7**, 2591-2601 (2014)
7. M. Amani, B. Salehi, S. Mahdavi, B. Brisco, ISPRS J. Photogramm., **144**, 119-136 (2018)
8. Y. Duan, X. Huang, J. Xiong, Y. Zhang, B. Wang, Int. J. Digit. Earth, **9**, 851-872 (2016)
9. M. Wang, Z. Zhang, Y. Zhu, Z. Dong, Y. Li, J. Real-Time Image Pr., **15**, 565-581 (2018)
10. O.L. Makarov, D.K. Mozgovoy, V.S. Khoroshilov, 21th Int. Conf. on Microwave and Telecommunication Technology (CriMiCo), 905-907 (2011)

NON-PARAMETRIC APPROACH FOR PRELIMINARY PROCESSING OF EARTH REMOTE SENSING DATA

Maksim A Denisov[1] , *Ekaterina* A Chzhan[1] and *Anna* A Korneeva[1]

[1]SibFU Institute of Space and Information Technologies, Siberian Federal University, 660074 Krasnoyarsk, Russian Federation

Abstract. Researches described in the paper are aimed at studying the methods of data preprocessing from a sample of observations of a system characterized by input-output values of variables. We consider the data containing omissions and outliers. Algorithms for leveling outliers in a sample of observations, as well as algorithms for filling data gaps are presented. In addition, it is implemented a data repair algorithm that is able to recover lost values (outliers) after their exclusion. Our studies are useful in geographic information systems or in the analysis of information received from satellites during remote sensing of the earth.

1 Introduction

The sample of the input-output observations of the system (object) under study plays a valuable role in solving the problem of identification. The data in the tables with the values of observations of the object can be both quantitative and qualitative. Qualitative data can be, for example, geographical name, species composition of vegetation, soil characteristics, etc. Paper [1] describes how to work with such type of data obtained from satellite sensors. Author in paper [2] explores the potential of using geographical information systems (GIS) and paper [3] considers the applied side of working with quality data in such systems. In this article, the sample is presented in the form of numbers – quantitative data type.

Most often, in practical problems, the data may contain defects of various kinds: outliers or omissions.

Omissions arise, for example, during the process of shooting or transmitting data from satellites in cases when the transfer process was interrupted or simply due to some technical malfunction of the shooting device. All of this may interfere with the further processing of the image. Algorithms for filling such gaps using nuclear functions are developed under the condition of parametric identification [4], which differs from the approach proposed in this article.

The occurrence of outliers in a sample of observations may be due to an inaccurate mathematical model, malfunction or improper calibration of instruments, mistaken readings, gross recording, and calculation and execution faults. The impact of outliers can be leveled using the robust identification algorithms described in books [5, 6] and papers [7, 8] or using data censoring methods, one of which is described in [9]. In particular, in [10] it is described outliers detection in the field of hyperspectral imagery which is connected with the task of locating pixels with spectral signatures that deviate significantly from the local background. As a result, such significantly deviate values in data affects the final accuracy of the object of study approximation.

The paper discusses two methods of data preprocessing, where the first one is a method of censoring a sample of observations to remove outliers, and the second one is restoring omissions using a non-parametric identification algorithm. Previously, the analysis of these algorithms cumulatively was not considered, which underlines the relevance of this work. In addition, after removing the outliers, it was decided to review and implement the data repair algorithm, which additionally increase the accuracy of object modeling.

2 The problem statement

An object of a discrete-continuous type is examined, whose scheme in its general form is shown in the figure below:

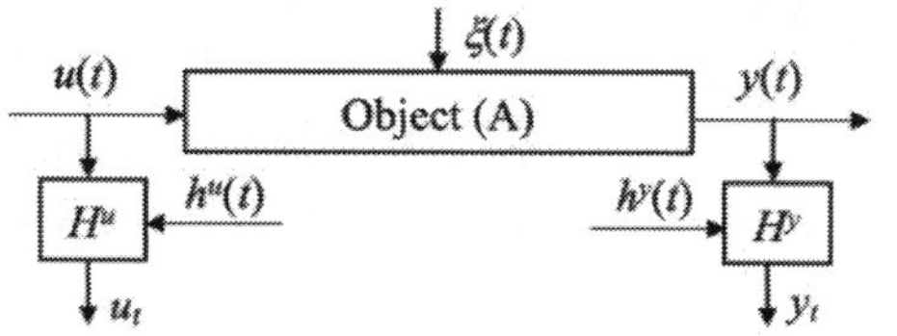

Fig. 1. Discrete-continuous process flowchart

The notation of the figure 1: A is unknown object functional; $y(t)\in\Omega(y)\subset R^1$ is an output process variable; $u(t)=(u_1(t), u_2(t),\ldots,u_m(t))\in\Omega(u)\subset R^m$ is a vector of input signal, where m is the number of input signals; $\xi(t)$ is the vector random noise; t is the continuous time; H^u, H^y are communication channels; $h^u(t)$, $h^y(t)$ are stochastic noise measurements; $\{u_{ji}, y_i, i=1,\ldots,s, j=1,\ldots m\}$ is a training sample, where s is sample size.

Most often, priori information about the object of study is not sufficient to build a model with an accuracy of the vector of parameters (in other words it is difficult or impossible to built a parametric model), therefore, a non-parametric approach is used. In this paper, the Nadaraya-Watson non-parametric estimation is used for conducting computational experiments. It is described by the following equation:

$$\hat{y}(u)=\sum_{i=1}^{s} y_i \prod_{j=1}^{m} \Phi\left(\frac{u_j-u_{ji}}{c_s}\right) \Bigg/ \sum_{i=1}^{s} \prod_{j=1}^{m} \Phi\left(\frac{u_j-u_{ji}}{c_s}\right), \quad (1)$$

where c_s – bandwidth parameter, and $\Phi[(u_j-u_{ji})/c_s]$ is a kernel function.

Due to the fact that we do not have a sample of observations with outliers and omissions, it is necessary to generate it by yourself as part of a computational experiment.

Omissions are generated as follows: $t=3\Delta t$, where t is the time step. Let us suppose, there is a sample with the volume s = 300. In order to obtain a sample with omissions according to the method described above, it is supposed to leave every third value in the data table. Thus, for a sample containing, for instance, s = 300 objects, omissions equal 200 values but other 100 values form remaining sample of observations.

Below, in Table 1, the initial sample is presented in general form (a "clean" sample without outliers or omissions); Table 2 contains a sample with omissions generated according to the rule described above.

Table 1. Initial sample

u_1	u_2	...	u_n	y
u_{11}	u_{12}	...	u_{1n}	y_1
u_{21}	u_{22}	...	u_{2n}	y_2
u_{31}	u_{32}	...	u_{3n}	y_3
u_{41}	u_{42}	...	u_{4n}	y_4
u_{51}	u_{52}	...	u_{5n}	y_5
...	...	...	...	...
u_{s1}	u_{s2}	...	u_{sn}	y_s

Table 2. Sample with omissions

	u_2	...	u_n	y
u_{11}	u_{12}	...	u_{1n}	y_1
u_{21}	u_{22}	...	u_{2n}	–
u_{31}	u_{32}	...	u_{3n}	–
u_{41}	u_{42}	...	u_{4n}	y_4
u_{51}	u_{52}	...	u_{5n}	–
...	...	...	...	...
u_{s1}	u_{s2}	...	u_{sn}	y_s

Outliers are generated for the remaining sample values (y_1, y_4, ..., y_s from Table 2). The value number that will be an outlier is a pseudo-random. After the number has been selected, the value in the sample is changed by the following formula:

$$y_i = y_i \cdot k \cdot c \qquad i \in [1; s], \tag{2}$$

where $k = 0.99$ is noise coefficient and c is normally distributed random value in the interval [-1;1].

3 Algorithm for filling omissions in data

Filling the missing values in the observation matrix is performed using nonparametric estimation (1). The application restoration algorithm of this type for objects of different mathematical description is considered in [11].

At the first stage, non-parametric identification of the object of study is performed using (1) from a sample of observations, in which omissions were removed in advance (cells with gaps from Table 2), that is a sample: $\{u_{ji}, y_i, i=1,\ldots,s_2; j=1,..,m\}$, where $s_2<s$ is sample size after removing all omissions. Next, the gaps in the observation matrix are filled using the estimate (1) obtained in the previous step. In those cases, where observations of the output of the object y are omitted, the known values of the input of the object u are substituted into the estimate. Thus, the values of the omissions are restored from the available input data about the object of study. The formula for calculating missing values is presented below:

$$\hat{y}_1(u) = \sum_{i=1}^{s-s_2} y_i \prod_{j=1}^{m} \Phi\left(\frac{u_j - u_{ji}}{c_s}\right) \Bigg/ \sum_{i=1}^{s-s_2} \prod_{j=1}^{m} \Phi\left(\frac{u_j - u_{ji}}{c_s}\right), \tag{3}$$

where u – the value of the input variables for which the omissions of the object's output is restored. As a result, the nonparametric estimate (1) is calculated anew for the elements of the reduced matrix.

4 Data censoring algorithm

As a method for eliminating outliers from a sample of observations, a data censoring algorithm will be used.

In the first stage using a sample of observations without gaps $\{u_{ji}, y_i, i=1,\ldots,s_2; j=1,..,m\}$ a non-parametric model of the form (1) is constructed. After that, the following condition is checked for all sample values:

$$|y_i - \hat{y}_i| > \mu \cdot \Delta, \tag{4}$$

where y_i is output values of the object, $\hat{y}_i$ is output values of the model (1), μ is customizable parameter, Δ is parameter, whose value is defined as:

$$\Delta = s^{-1}\sum_{i=1}^{s}\left|y_i - \hat{y}_i\right|. \tag{5}$$

If a sample point satisfies condition (4), then it is marked as an outlier and is removed from the sample of observations. As a result, the new sample is of the form: $\{u_{ji}, y_i, i=1,\ldots,s_3; j=1,..,m\}$, where $s_3 < s_2$ – volume of censored sample.

After applying the operations described above, the non-parametric non-outliers model is reconstructed from s_3-volumed sample of observations, the mathematical description of which is presented below:

$$\hat{y}_2(u) = \sum_{i=1}^{s_3} y_i \prod_{j=1}^{m} \Phi\left(\frac{u_j - u_{ji}}{c_s}\right) \Bigg/ \sum_{i=1}^{s_3}\prod_{j=1}^{m} \Phi\left(\frac{u_j - u_{ji}}{c_s}\right). \tag{6}$$

Note that the number of outliers found depends on the parameter μ set by the user. The result of the most accurate detection of an outlier value will depend on how this parameter has been configured.

5 Data repair

Outliers exclusion from a sample of observations increases the final accuracy of the object approximation, but some of the data after censoring is lost. It is known, as a sample size decreases, the accuracy of modeling declines. In this regard, it was decided to repair the data. The term "data repair" refers to the identification and subsequent replacement of rough measurements (outliers) with values of the robust model [12].

The description of repair algorithm is as follows: the values that were labeled using the data censoring algorithm as outliers will be replaced with values of the non-parametric model (1).

6 Computational experiment

Let us assume that the mathematical description of the object under investigation is as follows:

$$y = 0.25u_1^2 - 0.25u_2^2 + \xi, \tag{7}$$

where ξ is noise imposed on the output of an object that is generated as:

$$\xi = y \cdot c \cdot k, \tag{8}$$

where c is normally distributed random value in the interval [-1;1] and k is noise coefficient.

The model will be calculated using the non-parametric estimate which is represented by the following mathematical expression:

$$\hat{y}(u) = \sum_{i=1}^{s} y_i \prod_{j=1}^{2} \Phi\left(\frac{u_j - u_{ji}}{c_s}\right) \Bigg/ \sum_{i=1}^{s}\prod_{j=1}^{2} \Phi\left(\frac{u_j - u_{ji}}{c_s}\right). \tag{9}$$

The kernel $\Phi[(u_j - u_{ji})/c_s]$ used in the estimate (9) has the form of a parabolic bell-shaped:

$$\Phi(z) = \begin{cases} 0.75 \cdot (1 - z^2), & |z| \le 1, \\ 0, & |z| > 1, \end{cases} \tag{10}$$

where $z = (u_j - u_{ji})c_s^{-1}$.

The bandwidth c_s is configuring using cross-validation based on the minimum of the quadratic criterion:

$$F = s^{-1} \sum_{i=1}^{s} (y_i - \hat{y}_i)^2 \to \min_{c_s} . \tag{11}$$

It is accepted a relative approximation error in the paper, which evaluates the accuracy of object modeling:

$$W = \sqrt{(s-1)\sum_{i=1}^{s}(y_i - \hat{y}_i)^2 \Big/ s\sum_{i=1}^{s}(\hat{m}_y - y_i)^2} . \tag{12}$$

At the first stage of the computational experiment, we generate a sample consists of s = 300 observations $\{u_{ji}, y_i, i=1,\dots,s; j=1,2\}$ using (7) with noise coefficient equals 5%. Further, create the gaps and outliers using the methods described above.

At the first stage of the computational experiment, we remove the observations containing omissions in the data and leave the outliers. Thus, we obtain a new sample of observations with a volume of s_2 = 100, after that we construct a non-parametric model (9). The accuracy of the modelling calculates using (12).

Next, filling the omissions using a mathematical expression (3). Combine them with the rest of the sample and get the original data volume s = 300. After that, we construct a non-parametric model (9), estimate the accuracy of which by (12).

In the next experiment, it is proposed to censor the data according to the method described earlier with a sample of volume s_2. The model (6) is constructed from a sample of volume s_3 (obtained after censoring), the accuracy is estimated using (12). The penultimate computational involves constructing a model (9) using a censored sample with filled gaps.

Finally, at the last stage of the computational experiment, for a censored sample with filled gaps, we apply the procedure of data repairing according to the algorithm described earlier. Thus, after conducting the entire series of experiments, the following results were obtained.

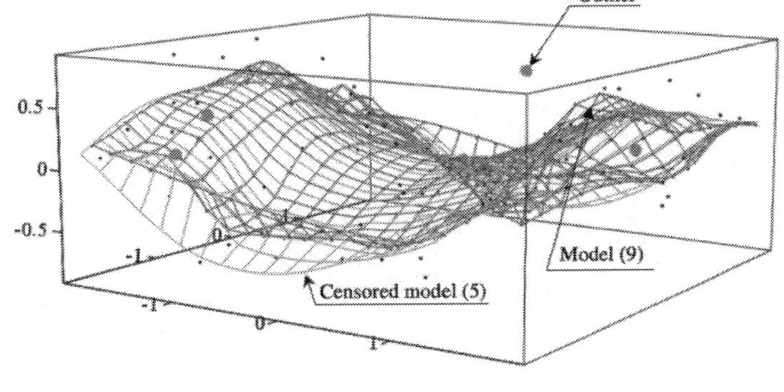

Fig. 2. Surface object (6) graph with omissions removed, model (9) and censored model (5)

Figure 2 shows the results of the computational experiment, when the omissions from the data were removed, after which, the data were censored of outliers. The graph shows that the censored model approximates the object more accurately than the model with outliers.

Table 3. The values of the relative error of approximation for all computational experiments

W_1	W_2	W_3	W_4	W_5
58.5%	35,2%	28,2%	16,6%	16,9%

In table 3, the following notation:
W_1 – relative approximation error for a sample with outliers and removed omissions;
W_2 – relative approximation error for a sample with outliers and filled omissions;
W_3 – relative approximation error for a censored sample with removed omissions;
W_4 – relative approximation error for a censored sample filled omissions;
W_5 – relative approximation error for repaired and censored sample with filled omissions.

Based on the results presented in the table above, it can be seen that when we fill omissions, as well as censor data, the accuracy of approximation increases by approximately one and a half to two times, which confirms the efficiency of the algorithms described in the work. In addition, the data repair algorithm after censoring also somewhat increases the accuracy of the final approximation of the object with the model.

7 Conclusions

The effectiveness of the algorithms described in the work confirmed an increase in the accuracy of approximation in the process of identifying the object of study. The computational experiment showed that sampling censoring gives more accurate simulation results than filling omissions in the data, however, if both of these methods are used together, then the final approximation results show the highest accuracy.

References

1. J. A. Engel-Cox, C. H. Holloman, B. W. Coutant, R. M. Hoff, Atmospheric Environment **38**, 2495-2509 (2004)
2. M. Pavlovskaya, Environment and Planning A **38**, 2003-2020 (2006)
3. J. Cidell, Area **42**, 514-523 (2010)
4. E. A. Lupyan *et al.*, Sovr. Probl. DZZ Kosm. (in Russ) **8**, 190-198 (2011)
5. P. J. Huber, *Robust Statistics* (2011)
6. R. Maronna, *Robust Statistics: Theory and Methods* (2006)
7. W. Zhong, H. Lu, M. H. Yang, *Computer vision and pattern recognition (CVPR)* 1838-1845 (2012)
8. T. Ding, M. Zhang, D. He, *International Conference in Communications, Signal Processing, and Systems* 1229-1236 (2017)
9. C. E. Brodley, Journal of artificial intelligence research **11**, 131-167 (1999)
10. G. Camps-Valls, L. Bruzzone *Kernel methods for remote sensing data analysis* 170-172 (2009)
11. A. A. Korneeva, N. A. Sergeeva, Vestnik SibGAU (in Russ.) **45** 49-54 (2012)
12. E. S. Kirik, Vychislitelnye tehnologii (in Russ.) **6** 351-355 (2001)

Experience of the application of remote sensing data in the problems of geo-ecological monitoring of the coal mining area

Olga Giniyatullina, *Evgeniy* Schastlivtsev, and *Vladimir* Kovalev

Institute of Computational Technologies of SB RAS, Kemerovo, Russia

Abstract. The experience of solving problems of geoecological monitoring of coal mining region with the use of remote sensing data is presented. The results of control over the boundaries of coal-mining enterprises, assessment of the degree of self-growth of dumps, monitoring of the state of vegetation near objects of coal mining and dust load of the area are shown.

The current state of the environment in areas with high anthropogenic impact requires constant monitoring by special services and the use of active tools for ecological monitoring. This problem is especially acute in coal-mining regions where as a result of mining fundamental changes in the structure of lands, landscapes, aquatic environments, biological communities, etc. occur. Classical methods of assessing the quality of the environment, for example, sampling and its analysis in the laboratory give the characteristics of objects only at a specific point while real changes can capture large areas that can far exceed the sanitary zones of mining enterprises. The use of remote sensing data to solve a number of problems of geoecological monitoring makes it possible to assess the state of the territories of the region as a whole. It should be noted that in this case laboratory test data become an authoritative source of results confirmation of space images processing, thus ensuring their high reliability.

In Kemerovo Scientific Center of ICT SB RAS for more than 10 years new approaches have been developed for solving geoecological monitoring problems using remote sensing data. The most successful developments that are applied in the work of a number of coal mining enterprises of Kuzbass are presented in this article. In general, remote monitoring is conducted in four directions:

1. determination of the boundaries of disturbed lands;
2. monitoring of reclamation / restoration land process;
3. assessment of the state of biomass in the territory of the facility and outside its sanitary zone;
4. obtaining dust load zones.

Let us provide examples of solving similar problems from each direction.

One of the most urgent tasks is to control the boundaries of licensed coal mining sites or the so-called "problem of black diggers". Violation of official boundaries leads to huge financial losses in the regional budget. It should also be noted that in such places uncontrolled coal mining is being carried out which leads to serious changes in the ecological condition of the area as the necessary measures are taken to minimize the negative impact on the environment. There is no work to restore disturbed lands.

In this case, the solution of the problem is to isolate the boundaries of the object under study in high-resolution images (for example, Sentinel 2A data of resolution 10 m / pix) and correlate them with the official boundaries of license areas using standard GIS methods (Figure 1).

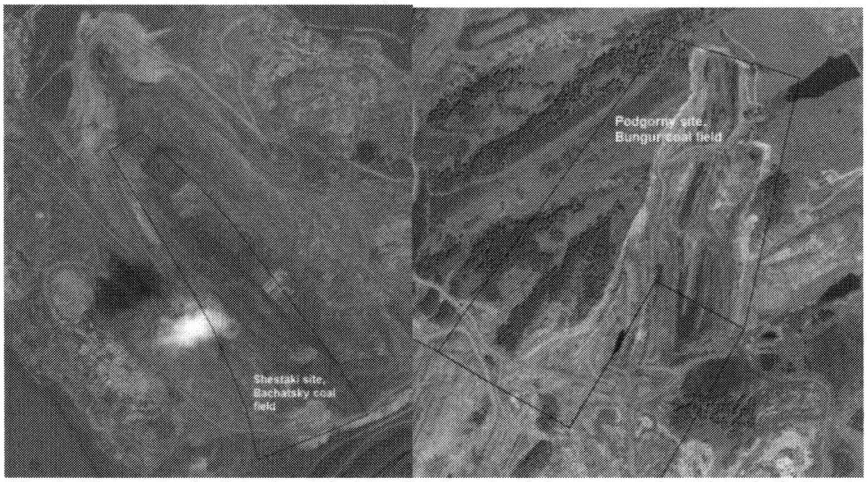

Fig. 1. The control of the boundaries of the coal-mining enterprise (the black line is the boundary of the license area)

As can be seen in Figure 1, the boundaries of both sites are partly outside the range indicated in the mining license. Therefore, an official letter of inquiry is sent to justify the increase in area. It should be noted that for example in this case the increase in the boundaries is justified by the technological need for coal mining in an open way. Transport infrastructure objects and tailing dumps and mine dumps are out of the boundaries, that is why that is not a violation and that enters the zone of the mining lands of coal-mining enterprises.

Another rather complex issue is the remote monitoring of land reclamation process after coal mining. The solution of this problem was described in details by the authors in [1]. The proposed method is based on the processing of a number of images over several years by the SAVI soil index from Landsat 7 data. The most sensitive spectral channel for counting this type of indices is the "extreme red" red-edge band (705-745 nm). The soil indices characterize the degree of coverage of the observed surfaces by vegetation and react well to the minimal presence of biomass on the surface of the object. From our point of view in this case the most convenient way for investigating the interrelated components of "vegetation-soil" is the use of the Gaudrian model. In this model the spherical coefficient of spectral reflection is used: it is assumed that the soil is a Lambert diffuser and the values of the coefficient do not depend on the direction of the sun's rays and sight incidence.

When using the SAVI index to estimate mine dumps the value $L = 1$ was used as objects with no or minimal vegetation expressed in the form of a weak grass cover were researched.

As a result of the processing, the changes in the state of biomass even in minimal amounts are recorded which is characteristic of self-growth processes when the appearance of vegetation is due not to the purposeful formation of the humus layer but to the chaotic transfer of plant seeds by natural processes. Thus, for example, in the study of the mine dump of the coal-mining section for the period from 1992 to 2010 when the work on its formation was actively conducted the indexes were fixed from -0.25 to 0.05, which indicates the appearance of biomass on the surface. Figure 2 shows the results of monitoring the mine dump of the Bachatskiy open-pit (Kemerovo region).

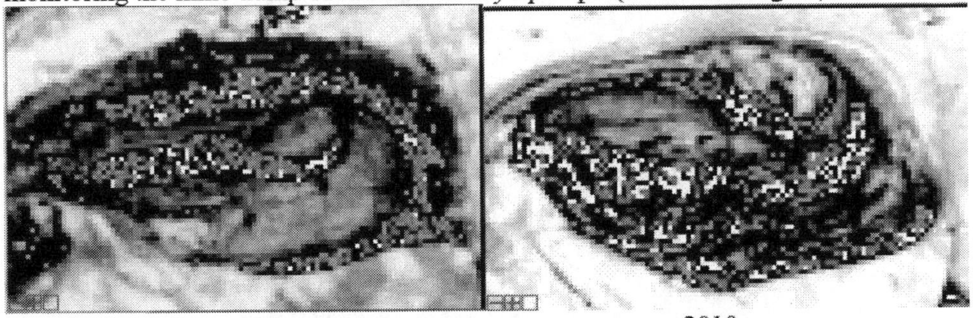

1992 2010

Fig. 2. Results of calculating the soil vegetation index SAVI

As can be seen, the shift of the horizon from the north to the south is clearly traced. The boundary of the soil index passes through 0, the color markers represent the indices in the range [0; 1] which characterize the soil. The indicators that are presented in white color fix inclusions which are an empty rock (i.e/ not a soil fertile layer). As it is clear from the figure the formation of the mine dump went from bottom to top and by 2010 an active process of self-growth of the southern side of the mine dump has been observed. It should be noted that reclamation of land in this mine dump in the specified period of time was not carried out while the average speed of appearance of the first stable manifestations of green mass is 4-5 years. While conducting the research on this method of experimental recultivation sites the index jump is fixed in the second year with its further increase depending on the prevailing species of vegetation.

A rather widespread problem is the estimation of biomass in satellite images using vegetation indices [2]. There are more than 160 different vegetation indices currently. The most common is NDVI (Normalized Difference Vegetation Index). In the course of research of different indices applicability it was proposed to use a combination of indices such as NDVI, Simple Ratio Index (SRI) and Anthocyanin Reflectance Index (ARI) to assess the state of vegetation in coal mining regions. If the first two indices are oriented to the allocation of biomass as a whole then the third index detects the intensity of anthocyanins in the leaves. Moreover, authors suggest to consider these indexes not separately but in combination. As a result, it is possible to obtain integrated vegetation maps based on joint processing of indices (Fig. 3) where the R band contains the values of the NDVI index, G-SRI and B-ARI.

04.06.2016 14.07.2016

30.08.2016

Fig. 3. The results of calculating the combination of NDVI, SRI and ARI indices

According to the accepted gradation, the normal state of vegetation is estimated in yellow-green tones which corresponds to the biomass of the research region. The presence of brown shades suggests a weakening of vegetation growth. However, as 2016 was very arid then quite large presence of this color is due to natural causes rather than the influence of coal mining. As a rule, weaker vegetation is fixed in the zones of direct contact with coal dump where the biomass is under considerable stress. But, as a rule, these changes are fixed in the zone not exceeding 500 m which corresponds to the established sanitary norms.

The next important direction of remote monitoring is the estimation of the dust load from the images over the winter period. The estimation is made by studying the intensity of the snow cover near the objects of coal mining. In [3] the authors described in detail the technique for processing images with object-oriented algorithms for selecting segments in an image. The essence of the approach lies in the complex isolation of structural, textural and spatial characteristics of objects in the image according to some rule or their combination. The approach is based on the law of emissions distribution from the source, i.e. the settling of pollutants occurs from the source of pollution with a decrease in concentration at a certain threshold when it is removed. As the maximum deviation in the IR zone is characteristic for snow with increasing pollution then the spectral brightness of its pixels can be determined and it is possible to divide them into some conditional classes in the presence of deviations in the IR spectrum. In fact, the task is to break down the image into segments using as a rule the deviation in the spectral values of the pixel in certain channels, i.e. internal properties of objects.

The best result for determining the distribution of dust on the snow cover is the use of an object-oriented algorithm with the setting of boundaries with varying intensity in any channel and using a complete list of λ values when small segments merge into larger ones. As a result, a segmented image of the snow cover is obtained where the boundaries of the change in reflection intensity of the coating are easily identified which characterizes the degree of its contamination.

It should be noted that in this technique the authors carried out a unique research of the confirmation of the results of image processing with calculated values from the ERA software package (www.logos-plus.ru) and data of laboratory researches of snow samples.

Figure 4 shows the results of processing a satellite image of Sentinel 2A using this method with the calculated values of the mathematical model.

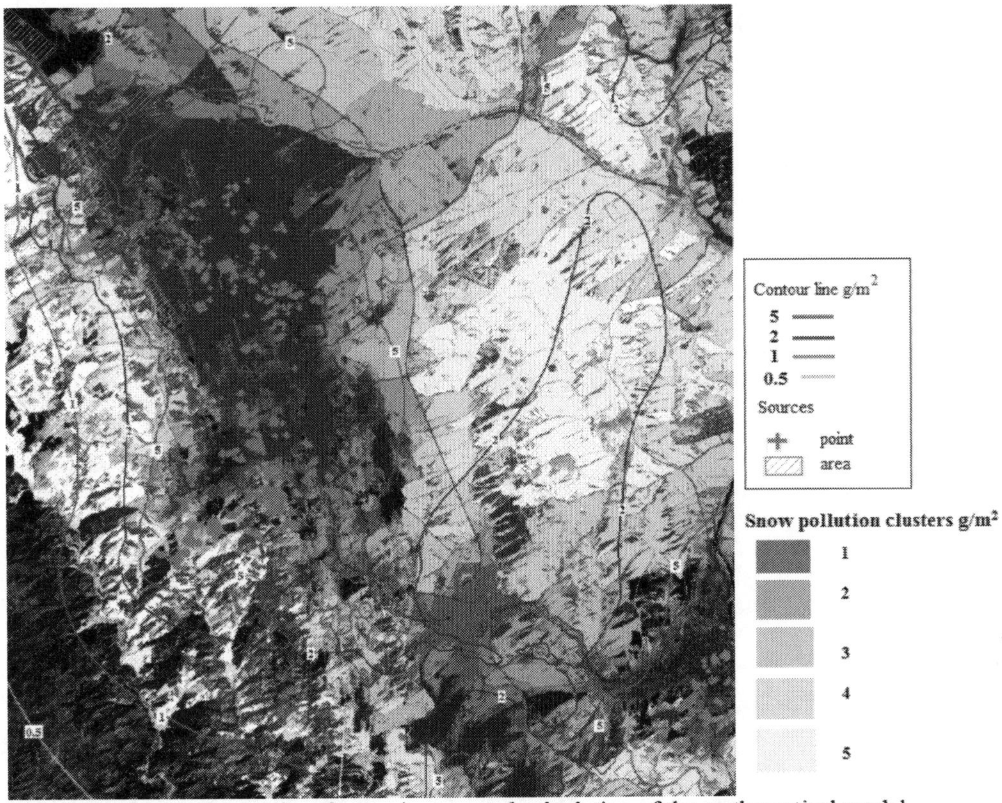

Fig. 4. Combining the results of space images and calculating of the mathematical model

As can be seen, there is a 100% hit of objects with a sharp drop in the reflection values in the region of the IR spectrum in the contour of the calculated contour line of 5 g /m2 (Fig. 4, line 5). The resulting segments correspond to the sources of contamination. After that, the allocated segments are distributed to four more classes with increasing values in the IR spectrum (Fig. 4, lines 1-4) which corresponds to the distribution of contaminators from sources. Thus, it became possible to identify the extent of snow contamination in the officially accepted classification.

As can be seen from the figure, the approach that was developed by the authors to image processing allows to obtain a topical picture of the dust load of the coal-mining region. The thematic products obtained on the basis of this approach make it possible to monitor the dust load of a region with a high technogenic impact throughout the region, and not only at ground control points.

Of course, the above list of tasks is not exhaustive. In this article we have only touched upon the most frequently solved problems of geoecological monitoring using remote sensing data for the coal-mining region. This list is being actively expanded. The emergence of new remote sensing data, methods and approaches to their processing allows creating fundamentally new thematic products and instruments for monitoring the state of the environment.

References

1. V.N. Oparin, V.P. Potapov, O.L. Giniyatullina, E.L. Shchastlivtsev, Studies into the process of mine waste dump filling up by vegetation using remote sensing data , JMS, Vol. 49, No. **6** (2013).

2. John G. Lyon, Ding Yuan, Ross S. Lunetta, Chris D. Elvidge, A Change Detection Experiment Using Vegetation Indices, Photogrammetric Engineering & Remote Sensing, Vol. 64, No. **2** (1998)
3. V. N. Oparin, V. P. Potapov, O. L. Giniyatullina, A. A. Bykov, and E. L. Schastlivtsev, Integrated monitoring of induced air pollution in mining regions, JMS, Vol. 53, No. **5** (2017)

Use of remote Earth sensing for agro-ecological estimation of soil cover structures and optimisation of agriculture in the forest-steppe zone of the Krasnoyarsk territory

Yuliya V. Gorbunova, *Tatyana* N. Demyanenko*, and *Valentina* V. Chuprova

Federal State Budget Educational Institution of Higher Education Krasnoyarsk State Agrarian University, 90 Mira Avenue, Krasnoyarsk City, 660049, Russia

Abstract. The possibilities of using data of Earth remote sensing for the construction of a geomorphological profile for the purpose of agro-ecological assessment of soil cover patterns and subsequent land typification as a basis for the design of adaptive landscape agriculture in the Krasnoyarsk Territory are considered. The geomorphological profile illustrates the features of the relief and regularities in the distribution of soil structures with a certain set of agro-ecological conditions and limiting factors that determine the choice of agro-technologies and farming systems.
Keywords: SRTM data, digital relief model, digital soil map, geomorphological profile (catena), structure of soil cover.

Introduction

Use of the Earth remote sensing (ERS) data is one of current trends in increasing efficiency of agricultural industry. It is caused by the ability of GIS software to solve the tasks of spatial location of cultivable areas with a set of crops corresponding to their fertility. The use of satellite data allows analyzing large areas, revealing migration trends, constructing sets of special maps by certain parameters which characterize various aspects of cultivated land. An outstanding feature of the obtained data is providing possibility for the development of new agro-technologies to optimize agriculture.

To use their landholdings efficiently, agricultural enterprises must be well aware of, first of all, the relief and the soil structure of cultivable areas [1,2]. Relief is the key factor of soil formation; it has a direct influence on the processes flowing in soil stratum, including development of erosion phenomena. Relief determines spatial heterogeneity of topsoil which often leads to allocating various types of soil even within the limits of one farm field. This is why ERS technology is used to solve technological tasks in agriculture. Catena transects, being commonly used as cartographic units, are characterized by the biggest certainty of detecting substance migration flow [3] and topsoil functioning related to it.

* Corresponding author: t-demyanen@mail.ru

The aim of this work is to create a geomorphological profile (catena) with the use of ERS data for agro-ecological estimation of soil cover structures and typification of lands.

Objects and methods

The cultivated land used by ZAO "Iskra", Uzhursky district, Krasnoyarsk territory, has been chosen for research. One field contains a geomorphological profile which represents a catena (centre coordinates: 89,76° and 55,40°) on a convex slope with a steepness about 4,5o covered by agricultural vegetation. Catena length is 2250 m. The catena ends with a massive degradation.

Three positions, or elementary geochemical landscapes, were allocated, samples of soil profiles were taken, and the main outlines of soil profiles are designated according to the relief conditions on the catena during the on-site inspection. The on-site inspection was preceded by the choice and creation of a geomorphological profile with the use of MapInfoProfessional software. SRTM data were used to explore the data [4]. SRTM (The Shuttle Radar Topography Mission) provides the relief data for almost all the Earth surface obtained by "Space shuttle" spacecraft to create the most complete electronic database of the high definition Earth relief data.

Basing on the space imagery results, the catena position has been worked out, large-scale thematic maps titled "Relief" and "Topsoil" have been compiled and superimposed in two layers. To digitize "Relief" map we used the soil map of the farm territory compiled in 1988 by Roszemproject of Krasnoyarsk Vostsibgiprozem Institute in scale 1:25000.

Results and discussion

ZAO "Iskra" is located in Chulym-Yenisey basin, restricted by the Batenevsky range (the Kuznetsk Alatau spur) in the south and by the Solgon range (the Eastern Sayan spur) in the north. On this territory the relief is considered to be formed under the influence of [5] glacial processes (displayed on the mountain summits), erosional processes (associated with the ground layer of mountains and ranges, accumulative when declining). A distinctive feature of the present-day relief of the basin is heterogeneity [6]. Figure 1 represents the digital relief model of the area. It is apparent that the distinctive relief forms of the southern part of the area are hilly uplifts, hummocks, and monoclinal hills (cuestas). Flat-plain relief of intercuesta degradation is noted in the central part, and hilly-plain relief with ridges and hills in the northern part. Slope location and "mountainous" framing (720-815 m. above sea level) of the area account for the development of denudation, diluvial, and erosional accumulative processes.

The climate of the area is very continental (Ivanov index of continentality – 210) and semihumid. According to the figures provided by the Uzhur weather station, the average annual precipitation equals to the average annual evapotranspiration, and it amounts to 464mm; the distribution of precipitation throughout the year is extremely inhomogeneous, only 17% of the annual precipitation occurs from November to April, the biggest amount of precipitation occurs in July and September (44% of the annual amount), often as showers which causes intense erosion on the slopes. Humidity factor of 1.0 characterizes the type of water regime as occasionally percolative. Selyaninov hydrothermal index is 1.3 which corresponds to sufficient humidity. Growing season is 154 days, but effective temperatures become stationary only on the 27th of May, thus the period of active vegetation amounts to 109 days.

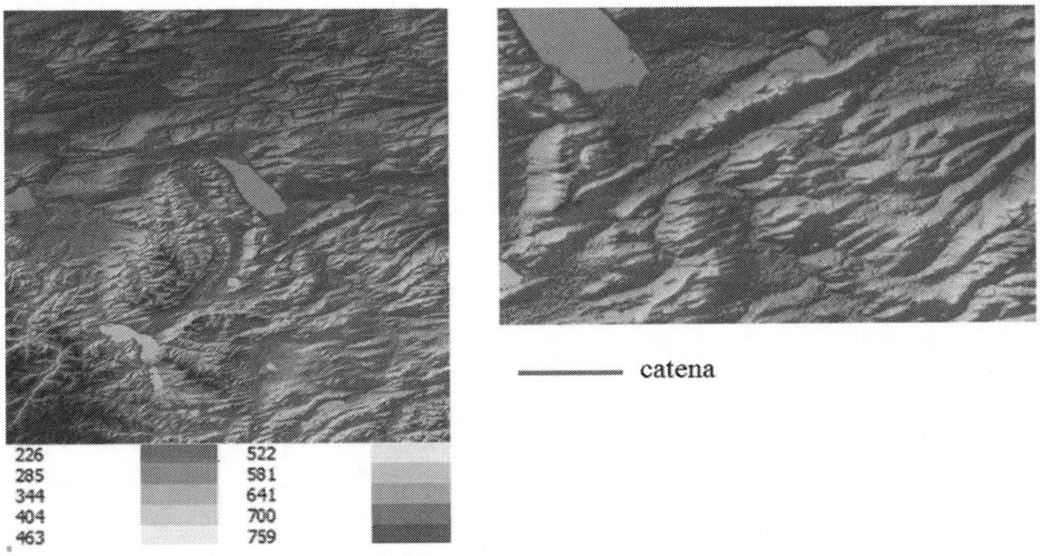

Figure 1 *– Digital relief model of ZAO "Iskra" area (map key - altitude above the sea level, m.)*

Cultivated lands are allocated according to the relief analysis, are differentiated depending on the soil parent material composition, ground water level, topsoil heterogeneity, physical and chemical soil characteristics, the degree of plowing of humic accumulative horizons, presence or absence of erosional processes. Positional distribution of cultivated lands is determined by the peculiarities of mesorelief and microrelief which widely vary over the different land areas of ZAO "Iskra".

Further, using the experience of field (on-site) inspection and certain experience of large-scale mapping of relief and topsoil structure, it is possible to specify allocated positions of elementary landscape, or even to divide it into additional categories. It is obvious now that the catena is remarkable for its complexity and accentuation both on lithological features and on migration flows which cause domination of denudation over erosion, or, vice versa, domination of erosional processes. More detailed division also implies compilation of special maps by certain parameters of migration processes.

The geochemical positions emphasized at this stage of the research are: eluvial denudation landscape (I), transitional landscape (II), eluvial accumulative landscape (III). They can be considered as microcatenas of a lower order (Fig. 2). The superimposed image of the layers titled "Relief" and "Soil map" of the cultivable area of ZAO "Iskra" (Fig.3) shows that soil-forming process runs under harsh conditions of rock weathering, short annual duration of soil-forming process, shallow invasion of soil weathering into sedimentation mass. The catena slope is characterized by bedding of soil parent materials close to the surface, as indicated by the abundance of coarse waste. Agricultural use of this territory causes intensive anthropogenesis which launches erosion accumulative processes. It is shown up in wide spread of shortened humus horizon soil and other erosional features.

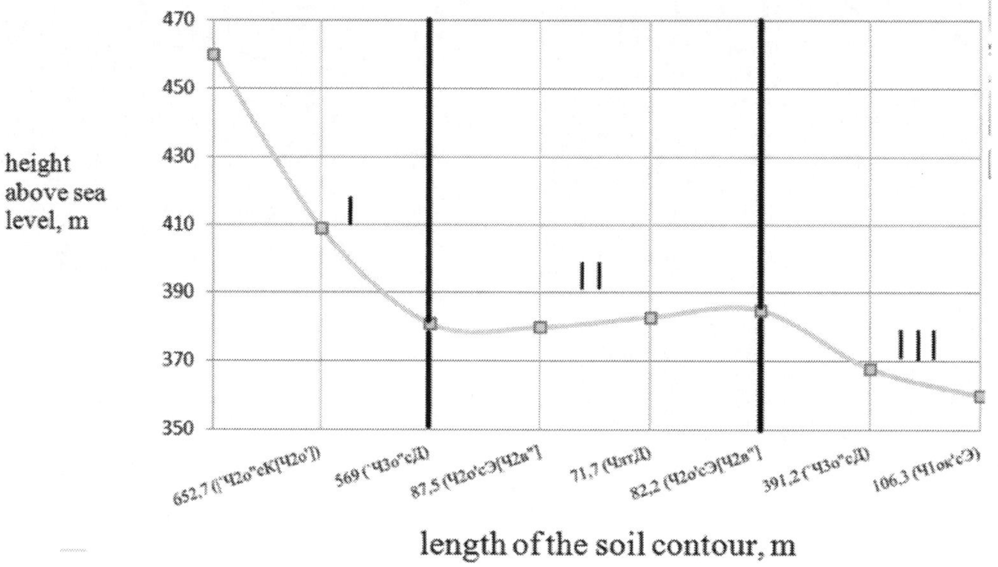

Figure 2 *– Geomorphological profile of ZAO "Iskra"*

Ч2о"сК[Ч2о'] – ordinary medium-humic, middle-deep, slightly eroded chernozem along with ordinary middle-deep chernozem (10-25%);

`Ч3о"сД – ordinary fertile, middle-deep, slightly eroded chernozem;

Ч2о'сЭ[Ч2в"] - ordinary medium-humic, shallow chernozem along with leached, medium-humic, middle-deep chernozem (10-25%);

ЧлтД - meadow chernozemic soil along with chernozemic meadow soil;

Ч1ок'сЭ - ordinary calcium low-humic, shallow chernozem,

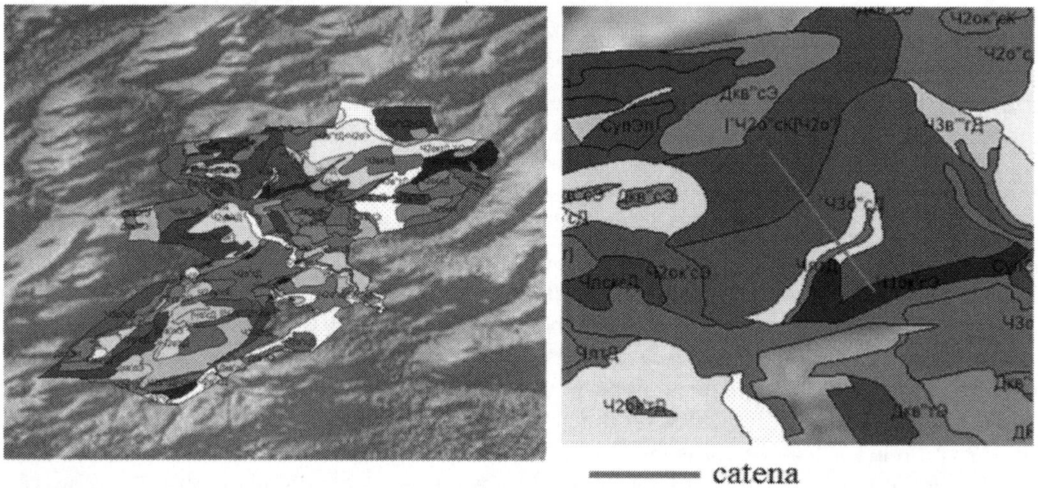

Figure 3 *– Superimposition of two layers "Relief" and "Soil Map" of ZAO "Iskra" cultivated area*

Lot I – the most elevational summit part of the catena with altitude marks from 380 to 460 metres above sea level; it has quite a high slope angle contributing to mainly horizontal material migration with atmospheric precipitation. This flow links the lot to the underlying catena positions. Vertical precipitation material inflow provides for the functioning of the soil profile of the ordinary shallow and middle-deep eroded chernozem complex. The top part of the soil profile is characterized by the high depth variability of the humus horizon and carbonate illuviation signs. The lower part of the profile is noted for its mottled red-brown colour, genetically bound to Devonian rocks. The texture of the whole soil profile is quite homogeneous, middle-loamy.

Lot II occupies the central part of the slope (380-385 metres above sea level), has a slight slope angle and is characterized by a gently concave surface with multiple microdepressions. Here atmospheric moistening prevails, erosional processes dominate, causing impoverishment of soil in humus and fertilizer elements. In depressed relief forms carbonate leaching is noticed. This is why the soil background is formed mainly by ordinary and 10-25%-leached types of chernozem. Land runoff development in this part of the catena is caused by the water flowing down from the surface of the overlying (position I) steep slope of the cultivated land.

Lot III of the geomorhological profile is characterized by a significant slope and ends in the valley part of the territory. Absolute elevation is 385-360 metres above sea level. Atmospheric moisturizing, often combined with ground water moisturizing. Mixed material migration continues here: vertical, influenced by ground water in microdepressions. Eluvial accumulative processes take place in this lot which causes bigger heterogeneity of the soil structure. The aggregate of automorphous and semiterrestrial regimes contributed to the formation of ordinary calcium shallow chernozems on local microhills and complex variations of ordinary deep meadow chernozemic and chernozemic meadow soils on microdepressions.

Thus, the soil massive of the catena is characterized by spatial heterogeneity of agro-ecological conditions and various fertility limiting features. The traditional approaches to agro-ecological estimation and land type grouping rely upon the studying and mapping of the factors which limit agriculture [8]. At this stage of the research only relief has been viewed as a factor of agro-ecological differentiation of the topsoil. The immediate task is to study indicational characteristics on various positions of the catena to develop digital maps of limiting features and agro-ecological types of soil.

Conclusion

As a conclusion, it should be pointed out that the use of the Earth remote sensing data opens up new opportunities of studying catena peculiarities of cultivated lands to estimate their agro-ecological state. Through the example of the catena of the cultivated land in Uzhursky district it has been proved that complex slope relief pattern leads to domination interchange of eluvial and accumulative processes on various parts of the cultivated area, being influenced by erosional processes and other fertility-limiting features. That is why in such areas it is necessary to allocate corresponding types of land and to take them into consideration when elaborating adaptive landscape farming systems, protecting and raising soil fruitfulness.

References

1. Vlasenko A.N., Dobrovolskaya N.I., Semendyaeva N.V. Agro-ecological estimation and typification of lands as the basic element for projecting adaptive landscape farming: a guideline, GNU SibNIIZHSH. – Novosibirsk, 2011, - 55 p.
2. Kudryashova S.Ya., Ditz L. Yu., Chichulin A. V., Chumbaev A.S., Miller G. F., Bezborodova A.N. Ecological - Geographical Aspects of Soil Complex Types Allocation at the Ukok Plateau Using Remote Sensing Studies // Siberian Ecological Journal, 2012, No. 5, pp. 703-710
3. Bogdanova M.D., Gerasimova M. I., Gavrilova I.P. An experience of large-scale mapping of the main objects of the geochemistry of landscape // Geochemistry of

landscapes and geography of soils. A report at the national conference, Moscow, 4-6 April, 2012. M.: Faculty of Geography MSU, 2012, pp. 64-66.
4. Onkov I.V. Accuracy estimation for SRTM altitudes for orthotransforming high definition space images // Geomatics. 2011, No.3. Pp. 40-46
5. Kallas E.V. Humic profiles of lake basins of Chulym-Yenisey basin. – Novosibirsk: Gumanitarnie nauki, 2004. – 170 p.
6. Tanzynbaev M.G. Soils of Khakassia. – Novosibirsk, 1993. – 300 p.
7. Weather archive in Uzhur [electronic resource] – URL: https://rp5.ru/Weather_archive_in_Uzhur
8. Sorokina N.V., Kozlov D.N. Methods of digital cartography in agro-ecological land estimation // Digital soil cartography: theoretical and experimental studies. – M.: V.V. Dokuchaev Soil Science Institute. 2012. – Pp. 140-154.

 https://doi.org/10.1051/e3sconf/20197502003

Management of the River Vedi Annual Runoff

Varduhi Margaryan[1,*], *Ksenia* Raevich[2]

[1]Yerevan State University, Department of Physical Geography and Hydrometeorology, Faculty of Geography and Geology, 0025 Alek Manoukian 1 Street, Yerevan, Armenia.

[2]Siberian Federal University, Institute of Space and Information Technologies, Chair of Systems of Artificial Intelligence, 660074 Kirenskogo St. 26, Krasnoyarsk, Russia.

Abstract. The article presents the following tasks for the Vedi river: to study and analyze the main physical and geographical factors determining the runoff; to collect, work out, analyze and estimate the results of factual hydro-meteorological observations of the river basin; to discover and analyze the features of intra-year distribution of the river runoff basin; to study dynamic changes of the river runoff; to create the methods of long-term forecasting of monthly and annual runoff; to create a map of spatial distribution of the river runoff. For these purposes were used the data of hydrological and meteorological stations, points of basin observations and mathematical-statistic, geographic, mapping, analytical and correlation methods.

1. Introduction

The Vedi river is the left tributary of the Araks, Fig. 1. The Vedi river basin is located on 800-2800 m height of Armenian volcanic plateu: the average sea level height is 2090 m. The length of the river is 58 km, up to Urtsadzor – 29 km, the square is accordingly 633 and 329 km2 in Table 1.

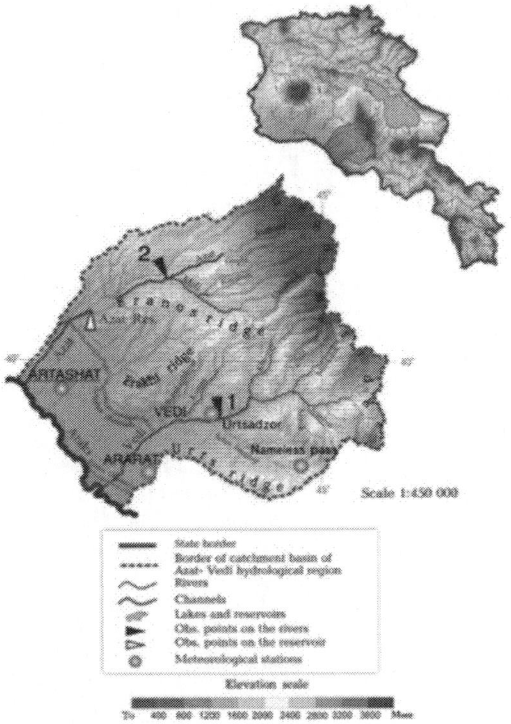

Fig. 1. The position of the Vedi catchment basin in the Azat-Vedi hydrological region
1-the r. Vedi – p. Urtsadzor; 2-the r. Azat– p. Garni

*Corresponding author: vmargaryan@ysu.am

Table 1. The main metrical characteristics of the Vedi river

River – point	From, km		River inclination, ‰		Catchment square	
	lade	sourc e	average	average balanced	square, km^2	average balanced height
The Vedi - v. Azizkend	39	20	66	60	142	2240
The Vedi – v. Urtsadzor	29	30	54	48	329	2090
The Vedi – v. Chimankend	25	33	51	45	348	2060
The SelavShahaplu – v. Shahaplu	10	-	-	-	59.0	-
The Kyusuz - v. Kyusuz	1,5	14	108	102	50.0	2260
The Akhsy - the lade	0,3	16	109	104	25.0	2020
The Khosrov – the pioneer camp	8,0	9,0	149	153	26.0	1980

The waters of the river are completely used for irrigation in the middle and lower flows. Therefore, the main significance lies in the studying of features of intra-year distribution of the river runoff, the estimation and management of the river runoff.

2. Tasks and solution methods

Taking into account all the above-mentioned, the following tasks have been solved:

- to study and analyze the main physical and geographical factors determining the runoff;
- to collect, work out, analyze and estimate the results of factual hydro-meteorological observations of the river basin;
- to discover and analyze the features of intra-year distribution of the river runoff basin;
- to study dynamic changes of the river runoff;
- to create the methods of long-term forecasting of monthly and annual runoff;
- to create a map of spatial distribution of the river runoff.

For these purposes the data of hydrological and meteorological stations and points of basin observations were collected, worked out and analyzed, which were kept in Armstatehydromet and used according to literary sources [1-5] and mathematical-statistic, geographic, mapping, analytical and correlation methods.

Flow formation is affected by permanently changeable meteorological factors and permanent physical and geographical factors. The features and regularities of the river runoff distribution are conditioned by the influence of physical and geographical factors. The main physical and geographical factors, which determine the characteristic of the river runoff formation, are geographical location, relief, geological and hydro-geological structure, climate and soil-vegetation cover of this area.

In the basin, the hydrological studies have been made since the 30s of 20th century. In the period of the study (1928 - 2009) 6 hydrological points worked in the basin, 4 of them worked only for one year (1968), and 1 of them for four years (1960 - 1963). There is only the Vedi river at the Urtsadzor hydrological point that is still working in the long series of observations. Although these observations started in 1928, the results of the observations have been kept since 1937, therefore, the data of 1937-2009 are used in the work. During its activity, the Urtsadzor hydrological point changed its former location several times: its catchment square, discontinuity of water level data. However, as the research found out, flows didn't have essential changes at the hydrological point in this period. Therefore, it is possible to use the results of water flow observations of 1937-2009 for forecasting and characterization of water regime basin.

In contrast to hydrological research in the basin, the meteorological researches have started later specifically since 1949. However, it should be noted that according to the continuity and homogeneity of meteorological observations the basin is one of the worst assimilated areas for forecasting. From the point of the river runoff assessment, the meteorological stations and points, which have rather long and reliable-sequence of data, are absent in the hydrological forecasts in the main zone of the stream flow formation. Thus, the data from the meteorological observations of stations in neighboring areas that have a relatively long data sequence are used to obtain forecast connections. It was found out that, when using the Artashat data, particularly good results are obtained for reconstructing the sequence of observational data at the Urzadzor and Ararat stations. They have got close correlations. However, using these data in forecast connections does not provide good results. Therefore, we used the actual data of observations from Artashat station in the work.

From the point of the river irrigation for the estimation of the river runoff, the important characteristic is intra-year runoff distribution by months and seasons of the year (table 2). It means, in terms of the water use, the intra-year runoff distribution and the water regime of rivers are important characteristics that depend on sources of supply and conditions of runoff generation and formation. The feeding source of any river plays an important role for intra-year runoff distribution and water regime. As the other rivers of the republic, the Vedialso river is fed with melted, rainy and underground water, which means that the Vedi river has the mixed feeding. In case of these conditions the Vedi river has an unstable water regime (large level fluctuations). For the river of Armenia, the Vedi river is also characterized with spring-summer floods, summer-autumn and winter stages of a low water level. As the other rivers of the Araks basin, the autumn-winter runoff of the Vedi river exceeds the summer-autumn runoff.

Table 2. Intra- year and seasonal distribution of long-term average values of runoff consumption vs. annual ones (%)

River – point	Months												Seasonal		
	I	II	III	IV	V	VI	VII	VIII	IX	X	XI	XII	IV - VI	IV - IX	X - III
The r. Vedi – p. Urtsadzor	4	4	6	20	36	13	4	2	2	3	4	3	69	77	23
The r. SelavShahap – p. Shahap	2	6	9	24	24	16	3	5	1	3	3	2	64	74	26

The annual runoff of the basin is characterized with one spring maximum (April-June). Moreover, a slight increase in consumption is observed during the autumn rains (fig. 2). However, it is important to note that the average monthly values of consumption during the spring flood can be 5-10 times more than the average annual values. In a low water level period the lowest monthly average values of consumption are 10-30 % from the average annual values. Even the drying of the river is sometimes observed and therefore it needs irrigation.

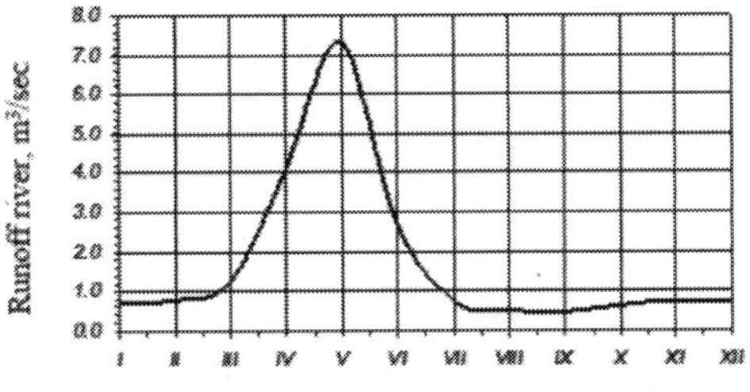

Fig. 2. Annual course of average values of runoff of the Vedi river at the Urtsadzor hydrological point

In intra-year runoff distribution, there is also some difference between cold and warm seasons of the year. In the warm period (IV-X), the Vedi river passes 80% of the total annual runoff and in the cold period (XI-III) – 20 %.

The average annual consumption according to the long-term at Urtsadzor hydrological point is 1,71 m3/sec. The maximum of the annual average value was observed in 1946 as 3,38 m3/sec while the minimum observed in 2004 was 0,37 m3/sec. From March the increase in water outlet is observed, which reaches to the maximum in May and beginning from June water outlets decrease almost in 2-3 times more and reach to the minimum in August-September.

During the year, runoff distribution increases the influence on the human economic activity more and more and therefore violates the natural state of water objects, consequently the runoff redistribution is observed. The water is taken from the river during the warm period for water supply purposes. Therefore, due to the intensive irrigation, water production is likely to decrease from July, and it is only 8% of the average annual volume in the period from July to September. However, natural runoff was not restored in the article.

After the floods there is a long-term shortage of water period with summer-autumn and winter stages in the river. In this stage the river is mainly fed with underground water and in winter it is fully fed like this. Nevertheless, the stage of water shortage sometimes (1965, 1979, 1982, 1989) observes disturbance in the summer-autumn periods because of floods after heavy showers. The least amount of water (once the river is not passed with the runoff and it becomes a dry hollow) is observed in the second part of summer unlike general regularity of runoff of the water period shortage in Armenia. It is conditioned by the anthropogenic factors due to the water supply point from the river, and as a result, it decreases the natural runoff of rivers. The average water supply point is 0,15-0,70 m3/sec.

As it can be seen, the river basin is characterized with unequal distribution of river runoff during the year. Therefore, it is necessary to organize correct regulation and management of the river runoff for economic productive water use of the river. The role of it increases more in conditions of climate change depending on dynamic change of the river runoff.

The scientific knowledge is necessary for more productive organization of regulation and management of runoff, which means, it should be forecasted. Thus, after studying, analyzing and estimation of physical and geographical conditions of the river basin, the literary sources have become the methods for long-term forecasting of monthly and annual runoffs. The monthly forecast is published on the 23-24 of the previous month.

After studying and analyzing of physical and geographical conditions, existing literary sources and initial materials have received some forecast dependencies, which can be applied during operational forecasting and consultation.

Stable winter regime makes it possible to establish a close connection between runoff values of neighboring months (previous and next). These dependencies are closely related and can be used for the forecasts of water runoff shortage. For the runoff forecast in January ($\overline{Q}_I$) and February ($\overline{Q}_{II}$) we obtained the following correlation equations:

$$\overline{Q}_I = 0.90\overline{Q}_{XII} + 0.12 \quad (1)$$

$$\overline{Q}_{II} = 0.90\overline{Q}_I + 0.13 \quad (2)$$

$$\overline{Q}_{II} = 0.66\overline{Q}_{XII} + 0.28\overline{Q}_I + 0.02 \quad (3)$$

In March, the water content is formed due to melting and underground runoff. The runoff in March ($\overline{Q}_{III}$) is conditioned by the water content in the previous period and the autumn-winter moistening level of the soil, which is expressed by the amount of precipitations.'

The forecast connections are based on the average monthly outlet values in January and February (($\overline{Q}_I$, $\overline{Q}_{II}$)) at Urtsadzor hydrological point and precipitation values from October to February at Artashat ($\sum X_{X-II}$) meteorological station.

$$\overline{Q}_{III} = 0.78\overline{Q}_I + 0.51\overline{Q}_{II} + 0.003\sum X_{X-II} - 0.04 \quad (4)$$

April, May and June are the months of flood period and they are relatively difficult to forecast. A large role in runoff formation is devoted to precipitations in the winter period, the amount of liquid precipitations of the flood period, the average air temperature, the summary of positive temperatures, and the runoff of the preceding period.

In April, the forecast connections are based on the average monthly outlet values in January, February and March at Urtsadzor hydrological point ($\overline{Q}_I$, $\overline{Q}_{II}$, $\overline{Q}_{III}$), precipitations of October-December period at Artashat meteorological station ($\sum X_{X-XII}$), air temperature in April ($\bar{t}_{IV}$), air moisture deficit in March ($\bar{d}_{III}$).

$$\overline{Q}_{IV} = 1.11\overline{Q}_{III} + 2.58\overline{Q}_{II} - 1.38\overline{Q}_I + 0.02\sum X_{X-XII} + 0.30\bar{t}_{IV} - 0.26\bar{d}_{III} - 2.27 \quad (5)$$

In May, as a rule, the main part of the annual runoff (35 %) flows out of the river. The water content in May is mainly formed with melting and partly with rainy waters. The average air temperature in April and the amount of precipitations within the preceding period were taken as the major indices, which characterize the residual moisture supply in the river basin.

The forecast connections are based on the average monthly runoff in March and April ($\overline{Q}_{III}$, $\overline{Q}_{IV}$), which are the absolute maximum, and the average air temperature in April at Artashat meteorological station ($t_{\max_{IV}}$, $\bar{t}_{IV}$):

$$\overline{Q}_V = 1.11\overline{Q}_{IV} + 1.12Q_{III} - 0.27t_{\max_{IV}} - 0.97\bar{t}_{IV} + 20.9 \quad (6)$$

In June, the forecast connections are based on the following values: the average monthly outlet in May ($\overline{Q}_V$) at Urtsadzor hydrological point, the average monthly and absolute maximum temperatures ($\bar{t}_V$, $t_{V\max}$), atmospheric precipitations in June (X_{VI}) at Artashat meteorological station:

$$\overline{Q}_{VI} = 0.36\overline{Q}_V - 0.57\bar{t}_V - 0.22t_{V\max} + 0.01X_{VI} + 16.3 \quad (7)$$

In July, the runoff is formed mainly due to underground water. Since the first decade of July the stage of water shortage is formed in the river. The precipitations, which fell on dry soil, did not flow into the river. Only in some years there was some increase of runoff when a large amount of precipitations fell on wet soil. The forecast connections are based on the average monthly values of outlet in June ($\overline{Q}_{VI}$) and air temperature in June ($\bar{t}_{VI}$) at Artashat meteorological station:

$$\overline{Q}_{VII} = 0.19\overline{Q}_{VI} + 0.21 \quad (8)$$

$$\overline{Q}_{VII} = 0.16\overline{Q}_{VI} - 0.08\bar{t}_{VI} + 2.09 \quad (9)$$

In August,in general, the runoff is formed due to the underground runoff component. The forecast connections are based on the average monthly values of runoff in July ($\overline{Q}_{VII}$),air humidity deficit ($\bar{d}_{VII}$) and wind speed ($\overline{V}_{VII}$) in July:

$$\overline{Q}_{VIII} = 0.48\overline{Q}_{VII} + 0.08 \quad (10)$$

$$\overline{Q}_{VIII} = 0.41\overline{Q}_{VII} - 0.02\bar{d}_{VII} + 0.12\overline{V}_{VII} + 0.39 \quad (11)$$

In September the runoff is formed due to the underground runoff component. The forecast connections are based on the values of outlet in August ($\overline{Q}_{VIII}$) and the average monthly wind speed in August ($\overline{V}_{VIII}$) at Artashat:

$$\overline{Q}_{IX} = 0.79\overline{Q}_{VIII} + 0.07 \quad (12)$$

$$\overline{Q}_{IX} = 0.64\overline{Q}_{VIII} + 0.15\overline{V}_{VIII} + 0.02 \quad (13)$$

In October the second maximum of precipitations is observed in the basin: its amount almost doubles compared to the amount of summer precipitations.

The forecast connections are based on the values of the average monthly outlets in August and September ($\overline{Q}_{VIII}$, $\overline{Q}_{IX}$), the air temperature in September at Artashat ($\overline{t}_{IX}$), the average monthly wind speed ($\overline{V}_{IX}$) and the precipitations in October (X_X):

$$\overline{Q}_X = 0.17\overline{Q}_{VIII} + 0.55\overline{Q}_{IX} - 0.04\overline{t}_{IX} + 0.10\overline{V}_{IX} + 0.01X_X + 0.82 \quad (14)$$

$$\overline{Q}_X = 0.69\overline{Q}_{IX} - 0.03\overline{t}_{IX} + 0.01X_X + 0.11\overline{V}_{IX} + 0.69 \quad (15)$$

$$\overline{Q}_X = 0.81\overline{Q}_{IX} + 0.21 \quad (16)$$

In November and December, the water content in the river is mainly formed due to the underground runoff.

The forecast connections are based on the values of the average outlet values in October ($\overline{Q}_X$):

$$\overline{Q}_{XI} = 0.95\overline{Q}_X - 0.15 \quad (17)$$

In December, the water content in the river is formed due to the underground runoff.

The forecast connections are based on the values of the average monthly outlet values in October and November ($\overline{Q}_X$, $\overline{Q}_{XI}$), the air humidity deficit in November ($\overline{d}_{XI}$) at Artashat:

$$\overline{Q}_{XII} = 0{,}77\overline{Q}_{XI} + 0{,}16 \quad (18)$$

$$\overline{Q}_{XII} = 0{,}06\overline{Q}_X + 0{,}72\overline{Q}_{XI} + 0{,}15 \quad (19)$$

$$\overline{Q}_{XII} = 0{,}06\overline{Q}_X + 0{,}75\overline{Q}_{XI} + 0{,}03\overline{d}_{XI} + 0{,}05 \quad (20)$$

Conclusion

Thus, we can observe the tendency of the river runoff reduction, especially since the 70s of XX century. It observes both of annual and maximum runoff.

There is a noticeable shortage of water period in the river with summer-autumn and winter periods: the minimum outlet of water is mainly observed in summer, which is conditioned by water supply points.

In the period of spring floods, the river passes 65-70 % of annual runoff. Moreover, the maximum expense of river water is often observed at this time. It usually occurs in April and in early May, but especially in the first half of May.

The forecasting methods, which were created as a result of the research, can be used for operational forecasting, and the river runoff can be managed with the help of the predicted values made in advance. We suggest accumulating water during the high water season for using it in a region with a low water level.

Practical estimation of hydrological forecasts, first of all, is determined by its accuracy and execution time. Therefore, in order to organize the hydrological service of that or another economic object, first of all, it is necessary to take into account the demand for time and forecast accuracy and find the optimal ratios between these two indices based on forecast possibilities of a particular hydrological phenomenon of a certain river basin.

References

1. Margaryan V.G. The Features of Formation of Araks Basins Rivers Runoff (in Armenia) and Regularities of Spatio-Temporal Distribution. Abstracts of the ISARM2010 International Conference "Transboundary, Aquifers Challenges and New Directions", Paris, France, 2010, p.173.
2. Margaryan V.G. The Features of Formation of Araks Basins Rivers Runoff (in Armenia) and Regularities of Spatio-Temporal Distribution. Abstracts of the ISARM2010 International Conference "Transboundary, Aquifers Challenges and New Directions", CD- ROM, Paris, France, 2010.
3. Margaryan V.G. The management of runoff of river Vedi based on long-term monthly forecasts. / Sovremennye problemy vodokhranilishh i ikh vodosborov. Gidro- i geodinamicheskie processy. Upravlenie vodnymi resursami. [Current problems of reservoirs and their water basins. Hydro- and geodynamic processes. Management of water resources], Mezhdunarodnaya nauchno-prakticheskaya konferenciya "Sovremennye problemy vodokhranilishh i ikh vodosborov" (29 – 31 May, 2015, Perm): in 2 volumes, Vol.1: Gidro- i geodinamicheskie processy. Upravlenie vodnymi resursami. / scientific supervisor A.B. Kitayev, O.V. Larchenko, - Perm: Perm state national research university, 2015. - p. 212-220.
4. Margaryan V.G. Assessment, management and regulation of the way of monthly runoff water season of Vedi river. Abstract book: "The international conference on regional climate CORDEX 2016", 17th - 20th May 2016, Aula Magna Stockholm University, Stokholm, Sweden. - P. 257.
5. Margaryan V.G. Estimation and Management of Annual Runoff of River Vedi. 7th International Water Resources Management Conference of ICWRS, 18–20 May 2016, Bochum, Germany, IWRM2016-17

E3S Web of Conferences **75**, 02004 (2019) https://doi.org/10.1051/e3sconf/20197502004
RPERS 2018

Anomalous atmospheric temperature perturbations over seismically active regions of Europe according to satellite measurements

Leonid Sverdlik[1,*], *Sanjar* Imashev[1], and *Tatiana* Yamskikh[2]

[1]Research Station of the Russian Academy of Sciences in Bishkek, Kyrgyzstan
[2]Siberian Federal University, Institute of Space and Information Technology, Russian Federation

Abstract. An algorithm for defining temperature anomalies and the results of diagnostic studies, based on a retrospective analysis of time series of satellite data in troposphere and stratosphere over the epicenters of strong earthquakes with magnitude M > 6.0, occurring in seismically active Mediterranean region of Europe are presented. It was found out that in all cases anomalous temperature disturbances were registered 1–10 days before the main seismic event.

1 Introduction

In recent years, there have been a number of studies, addressing the results of retrospective satellite remote sensing data analysis, which showed correlation between seismic activity and anomalous temperature variations in the upper troposphere during the catastrophic earthquake in Japan in spring 2011 [1], as well as on the eve of strong seismic events on the territory of Russia [2], in the northern Tien-Shan chains [3] and other seismically active regions of Asia from equator to mid-latitudes, both over land and ocean surfaces (Fig. 1) [4]. Despite the general positive statistics of temperature anomalies that preceded strong earthquakes, it was still impossible to state that these tropospheric effects are inherent to all major seismic events. Thus, there arose an obvious necessity for additional research, the main purpose of which, besides the development of methods for defining anomalous temperature perturbations in the tropopause, would be to enlarge the statistics of strong earthquakes. At the same time, we proceeded from the fact that analysis of connection between the dynamics of atmospheric and lithospheric processes, regardless of the method used, is based on the assumption that variations of parameters caused by seismic activity differ significantly from background fluctuations that occur during the periods when there were not recorded strong earthquakes, and, consequently, on allocation of areas connected with seismic activity in which a typical anomalous behavior in temporal data sequences using appropriate integral indicators was registered [5].

This paper describes the improved algorithm for defining short-period anomalous changes in the spatial-temporal distribution of temperature, and presents the results that were

* Corresponding author: l.sverdlik@mail.ru

obtained with the proposed algorithm application to strong earthquakes with magnitude M > 6.0 in the Mediterranean region of Europe.

2 Seismic and satellite data

Within the frameworks of this study, we analyzed results of atmospheric temperature satellite measurements on the eve of 8 strong earthquakes with magnitudes in the range 6.1 to 7.1, which were recorded in 2011–2016 in Turkey, Italy and Greece (Table 1 and Fig. 1). Seismic event data were obtained from interactive map service IRIS Earthquake Browser (IEB) (http://ds.iris.edu/).

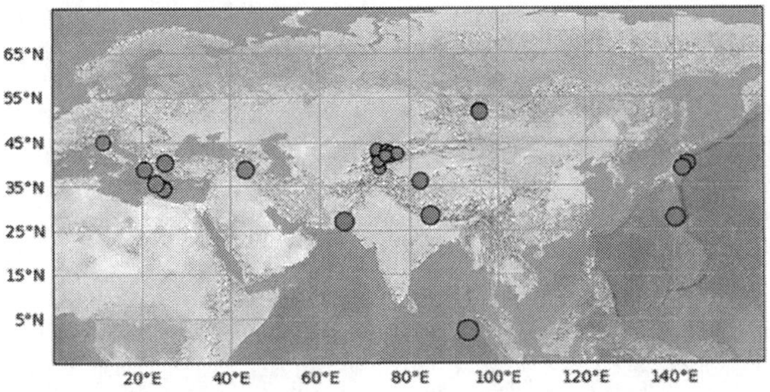

Fig. 1. The location of earthquakes epicenters.

Table 1. The features of seismic events under study (M > 6.0).

№	Date	Time	Latitude N	Longitude E	M	Region
EQ01	23.10.2011	10:41:22	38.72	43.51	7.1	Turkey
EQ02	20.05.2012	02:03:53	44.87	11.22	6.1	Italy
EQ03	15.06.2013	16:11:02	34.40	25.02	6.3	Greece, Crete
EQ04	16.06.2013	21:39:05	34.35	25.16	6.1	Greece, Crete
EQ05	12.10.2013	13:11:52	35.49	23.26	6.8	Greece, Crete
EQ06	24.05.2014	09:25:03	40.30	25.45	6.9	Aegean Sea
EQ07	17.11.2015	07:10:07	38.67	20.60	6.5	Greece
EQ08	30.10.2016	06:40:19	42.85	13.09	6.6	Italy

To analyze spatial and temporal variation of atmospheric temperature we used satellite measurements (AIRS / Aqua and MERRA-3D) for 10 pressure levels (p_k) varying from 500 to 70 hPa (http://giovanni.gsfc.nasa.gov/) covering the altitude from the upper troposphere to the lower stratosphere. The vertical profiles of the temperature *T(p)* were reconstructed over each earthquake epicenter (±1°) for time periods of at least 2–3 months before and after each of the EQ01–EQ08 events. The temporal discretization of initial temperature data *T(t)* was Δt = 3h.

3 Algorithm for processing the results of satellite measurements of atmospheric temperature

The algorithm for calculating the integral indicators of anomalous variations of temperature time series includes 6 main stages and is based on the use of a traditional approach, which consists in decomposing data into long and short-term components, with subsequent filtering and removing smooth trends.

3.1 Preprocessing of satellite data

At the first stage we formed continuous time series with different time resolutions (Δt = 3h, 12h and 24h) for each of the seismic events, containing temperature values at all altitude levels of atmosphere from 500 to 70 hPa. The duration of observation series $T(t)$ was 4–6 months. The height of the tropopause and cases of double tropopause were also determined on the basis of WMO criteria from temperature profiles $T(p)$ [6].

3.2 Spectral analysis of atmospheric temperature time series

Due to unsteady and non-linear nature of the factors influencing atmospheric temperature changes, large-scale (low-frequency) components of time series were detected using spectral analysis based on continuous wavelet transform. Using wavelet spectrograms we could reveal a hidden periodicity (quasi-permanent components), which must be eliminated along with low-frequency seasonal trend.

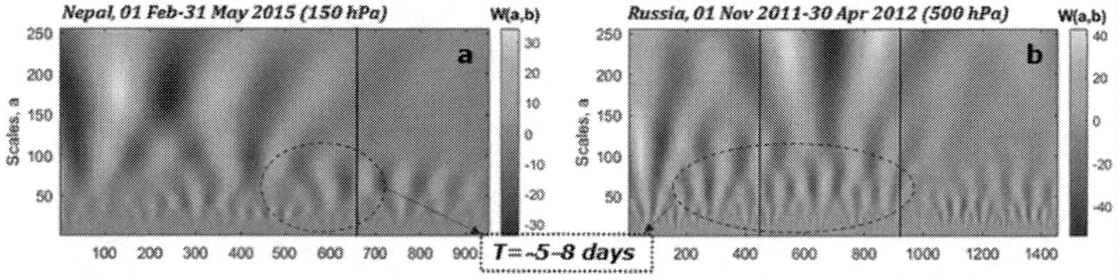

Fig. 2. Wavelet spectrograms of temperature time series (Δt = 3h) during periods of seismic activity in Nepal (150 hPa) (a) and Russia (500 hPa) (b). Vertical lines show the moments of strong earthquakes.

3.3 Allocation of short-term temperature variations

Processing of temperature time series is based on the analysis of quasi periodic oscillations with periods ~5–8 days. The validity of this approach is shown by wavelet spectrograms, clearly demonstrating variations of a similar time scale before strong earthquakes (Fig. 2). To allocate short-term variations we applied the nonlinear threshold-filtering method based on discrete wavelet transform [7], which has several advantages in comparison to classical methods, for example using moving average or higher-order polynomials.

3.4 Selection of height ranges

A fact-based selection of altitude ranges for carrying out calculations is very important for temperature data analysis. This procedure consists of determining the most informative pressure levels that are appropriate for specific locations, taking into account the features in the dynamics of thermal structure. For this, in each case we calculated correlation coefficients between temperature variations for each pair of levels (Fig. 3).

Temperature variations are non-uniform and differ significantly in separate atmospheric layers: the highest positive correlation is observed in the free troposphere, it decreases with increasing altitude to almost zero at the border of the troposphere and stratosphere and

 https://doi.org/10.1051/e3sconf/20197502004

changes sign in the zone of transition to positive gradients ($\gamma > \Delta T/\Delta p$). Square correlation matrix, which dimension corresponds to the number of the layers considered p_k, is a convenient way to represent the results visually. Based on such a matrix representation of the initial temperature data, it is possible to select reasonably the zones separated by the tropopause with the most intense antiphase oscillations (Fig. 3).

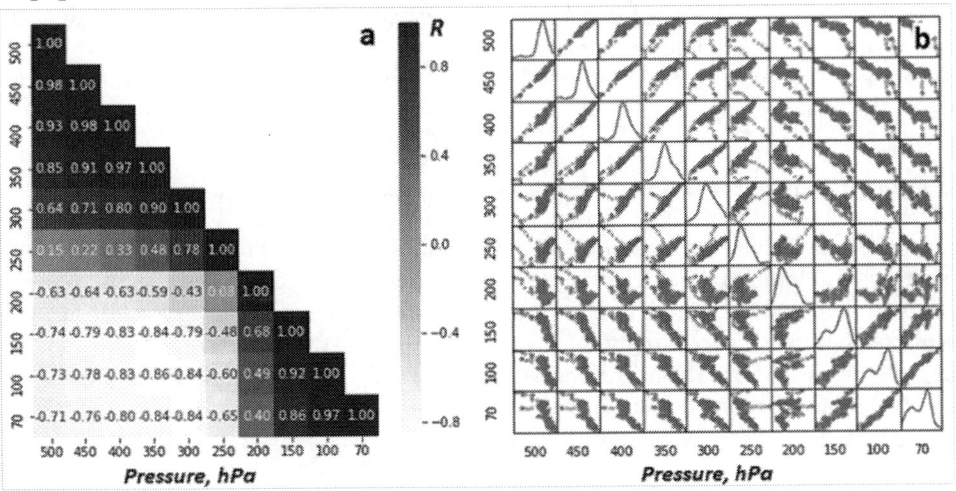

Fig. 3. Matrix representation of correlation coefficients between temperature variations on each pair of isobaric levels from 500 to 70 hPa.

3.5 Calculation of integral indicators for anomalous variations of temperature time series

We consider anomalous changes in amplitude and phase of temperature variations in the layers of atmosphere separated by tropopause as the main features characterizing temperature behavior of atmosphere before strong earthquakes (Fig. 4a).

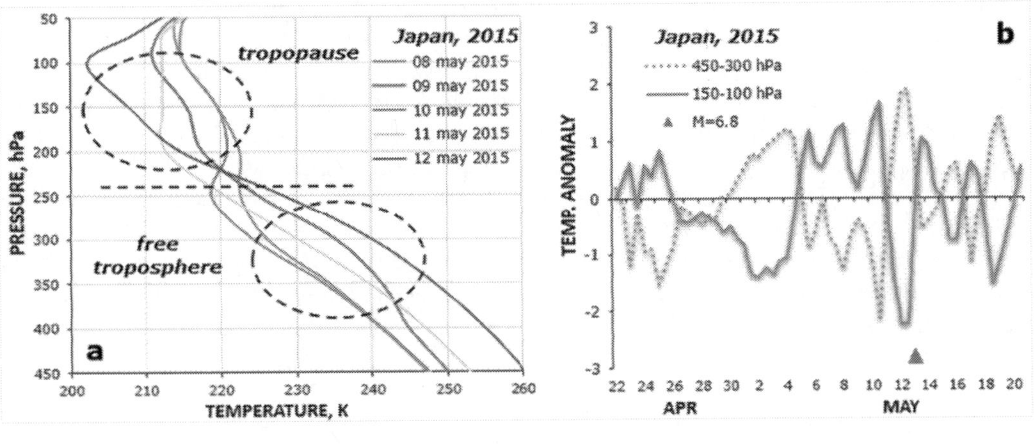

Fig. 4. Transformation of vertical temperature profiles (a) and time series of average values of temperature anomalies in 450–300 and 150–100 hPa layers (b) before the earthquake in Japan (marker indicates the moment of the earthquake $M = 6.8$ on May 12, 2015).

This feature was taken as a basis for developing an algorithm for detecting anomalous temperature perturbations [2]. However, until recently the product of moving variations of temperature anomalies time series was calculated irrespective of the phase of temperature fluctuations. Therefore, to improve the reliability of pre-seismic anomalies detection $T(t)$, the procedure for calculating D was corrected taking into account the correlation coefficient (R) between temperature changes at the considered levels or in the atmospheric layers, designated by D_{CORR}. We calculated the integral index D_{CORR} in accordance with the following conditions: $D_{CORR} = 0$, if $R \geq 0$ and $D_{CORR} = D \times |R|$, if $R < 0$. That is, the D_{CORR} parameter resets to zero at positive values of the correlation coefficient and is adjusted depending on the value of R with antiphase temperature changes (Fig. 4b). The inconsistency thus eliminated let us to diminish considerably although not to exclude the cases of greater than 2σ "false" (or in-phase) anomalies.

3.6 Analysis of anomalous values of D integral parameter and seismic data

The values of the product of temperature anomalies time series moving variations that are greater than one (D integral parameter) in free troposphere (ΘT_{FT}) and in tropopause (ΘT_{TP}) imply antiphase anomalies in short-term amplitude temperature variations in the considered layers of atmosphere, and, accordingly, high probability of seismic activity.

As experience has shown, the averaging of ΘT_{FT} and ΘT_{TP} for 2 or more isobaric levels is more preferable during calculations. This will provide additional "natural" data smoothing and selection of the most stable space-time variations.

4 Results of detecting temperature anomalies

Despite the amplitude and phase differences between temperature variations in the tropopause zone, determined by the seasonal conditions and geographic features of the considered earthquakes epicenters, implementation of the described algorithm made it possible to identify patterns (or features) common to all considered periods of seismic activity. A typical example of gradual temperature time series processing during foreshock activation and the main earthquake with magnitude $M = 6.1$, which occurred on May 20, 2012 in Italy, is shown in Fig. 5.

The curves of temperature changes at different levels do not provide insight into consistency with seismicity by themselves (Fig. 5a) or they are not obvious, although the non-random nature of the effect is often seen quite well. At the same time, the results of processing temperature measurement data, obtained in accordance with the algorithm for individual isobaric atmospheric levels in March – June 2012, showed distinct temperature anomalies, which were observed several days before strong earthquakes. As it can be seen, zones with the highest values of D parameter correlated with the main seismic impacts. The moments of anomalous temperature drop were quite short, and, which is the most important, they were directly connected to seismic shocks, preceded them (Fig. 5g). This feature of the atmospheric temperature behavior was tested for all selected cases of strong earthquakes in the Mediterranean region, thus confirming the results obtained previously [3].

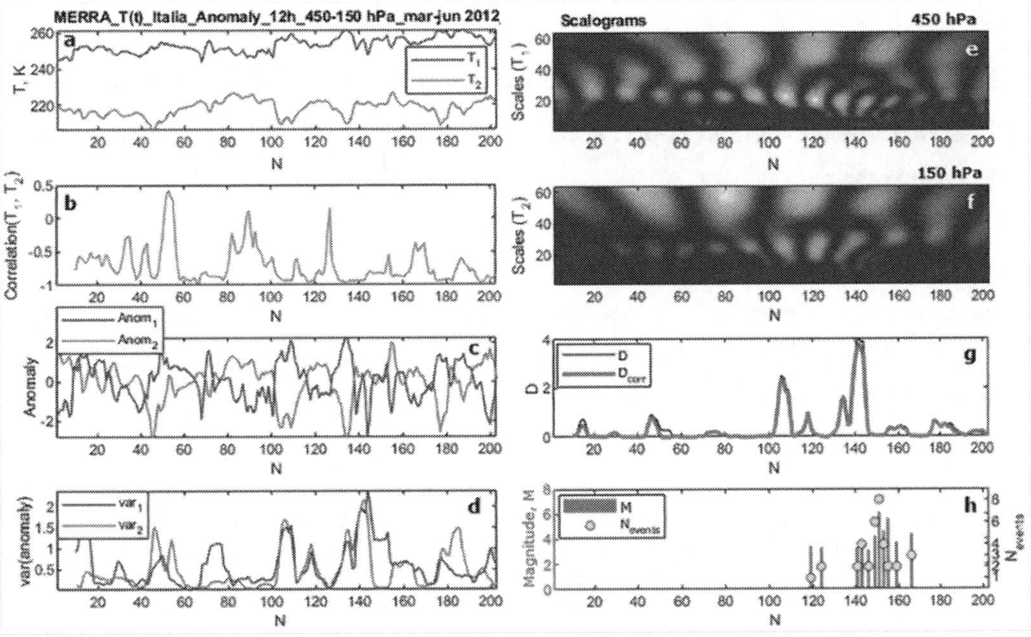

Fig. 5. Time series of temperature at levels of 450 and 150 hPa (a), correlation coefficient R (b), temperature anomalies (c) and their moving variations (d), wavelet coefficients spectra $W(a, b)$, (e, f) , D parameter *(g)*, magnitude M and number of earthquakes per day N (h) during the period from March to June 2012 (M = 6.1, 05.20.2012, Italy).

5 Conclusion

The algorithm for processing data of satellite remote sensing, developed and used in this study, made it possible to allocate anomalies of short-term temperature variations in the tropopause during the foreshock activation before strong seismic events.

An analysis of satellite data showed that strong seismic events with magnitude of $M > 6.0$ in the considered set that occurred in the Mediterranean region of Europe were preceded by an anomalous temperature change, which was observed about 1–10 days before the main shock and was followed by increase in the daily temperature range.

The results of remote sensing data processing revealed common features of anomalous temperature behavior synchronized with strong earthquakes in various regions of the world.

References

1. V.B. Kashkin. Atmospheric and Oceanic Optics, **26**(10), 908–916. (2013)
2. V.B. Kashkin, A.A. Romanov, A.S. Grigoriev, A.A. Baskova. J. Sib. Fed. Univ. Eng. Technol. **5**(2), 220–228. (2012)
3. L.G. Sverdlik, S.A. Imashev. J. Sib. Fed. Univ. Eng. Technol. **10**(6), 783–793. (2017)
4. L.G. Sverdlik, S.A. Imashev. J. Sib. Fed. Univ. Eng. Technol. **11**(8), 956–963. (2018)
5. M.R. Saradjian, M. Akhoondzadeh. Nat. Hazards Earth Syst. Sci. **11**, 1109–1108. (2011)
6. *World Meteorological Organization*. WMO Bulletin. **4**, 134. (1957)
7. V.M. Gadre, V.M. Dimri, E. Chandrasekhar. Boca Raton, FL: Taylor & Francis Inc. 286 p. (2014)

E3S Web of Conferences **75**, 02005 (2019) https://doi.org/10.1051/e3sconf/20197502005
RPERS 2018

Estimation of forest cover restoration damaged by Siberian silkmoth using remote sensing data and maps

Elena Fedotova[1,2]

[1]Sukachev Institute of Forest SB RAS, FRC KSC SB RAS, Akademgorodok,50, b. 28, Krasnoyarsk, 660036, Russian Federation
[2]Siberian Federal University, Institute of Space and Information Technology, 26, Kirensky st., Krasnoyarsk, 660074, Russian Federation

Abstract. The current state of the land cover has been estimated in the territories where in different years (1885, 1955, 1995) the forests were damaged by Siberian silkmoth. Dark-needle taiga is restored through the change of tree species. In 20 years in areas of dark-needle taiga there are graminoid communities, in 60 years we have deciduous forests there, and in 130 – dark needle forests, but not everywhere.

1 Introduction

The Siberian silkmoth is one of the most dangerous pests of the taiga forest, since during the outbreaks it damages forests in the area of hundreds of thousands of hectares until death [1]. This pest is a constantly acting factor in coniferous stands of Siberia, which are restored after its attack slowly and badly. The greatest damage from silkmoth is for forests with Siberian fir and cedar (Siberian pine), which are less resistant to damage and do not regenerate needles after it has been eaten by silkmoth.

The restoration of forest cover after the defeat by the Siberian silkmoth is an important topic for study, and this is due to the fact that forests have enormous ecological and economic importance. The existing works in this field is quite contradictory. So, N.F. Reimers [2] conducted observations of forest restoration after silkmoth outbreak in the territory of the Irkutsk region and Buryatia and came to the conclusion about the successful resumption of cedar (in the absence of repeated fires). According to the observations in the Tomsk region, V.A. Dudin (1958) came to the conclusion that forest restoration went through a change of tree species. This conclusion is supported by the materials of later studies in the forests of the Tomsk region by Pavlova (2004) and A.A. Ageeva (2006). Thus, either this process has regional peculiarities, or a variety of factors influencing the restoration of cedar after silkmoth outbreak it leads to different results.

The process of the forest cover restoration can be described as a succession. Succession is a gradual, irreversible, directed change of some biocenoses (a historically formed set of animals, plants, fungi and microorganisms inhabiting a relatively homogeneous living space and connected among themselves by their surroundings) by others on the same territory under the influence of natural factors or in the result of human activity [3]. It is of considerable interest to analyze the current state of vegetation cover in the areas where the outbreaks of the Siberian silkmoth took place at different times, which allows us to evaluate the course of restorative succession.

Succession after the Siberian silkmoth outbreaks is often complex. The recovery stages depend on the climatic conditions, the frequency of fires in a given area, the presence of undamaged forests as seed sources, human activity. In the area of the dead dark coniferous forests, as shown by numerous observations, after about 10 years there are shrubs and deciduous forests. Most often, after changing the tree species for 100 years or more, dark coniferous forest appears again on this territory.

The author's goal is to illustrate this concept of forest cover restoration for three Siberian silkmoth outbreaks that took place at different times in the territory of Siberia.

2 Study areas, Data and Methods

In this paper we consider three areas of interest (fig. 1) :

- the valleys of the Malaya and Bolshaya Belaya rivers of the Eastern Sayan mountains (the Irkutsk region) [4]; the silkmoth outbreak took place in the 1980s of the 19th century and damaged cedar (Siberian pine) forests;

- the Ket-Chulym interfluve; south taiga in West Siberia; the outbreak of 1952-1957 [1]; dark coniferous forests were damaged;

- the Lower Angara area (the Krasnoyarsk Territory); dark coniferous forests were damaged during the outbreak in 1993-1996.

The tasks of remote sensing data processing and their joint analysis with various maps were:

- to identify the territories of dead and damaged forests after outbreak;

- to determine the actual state of the land cover.

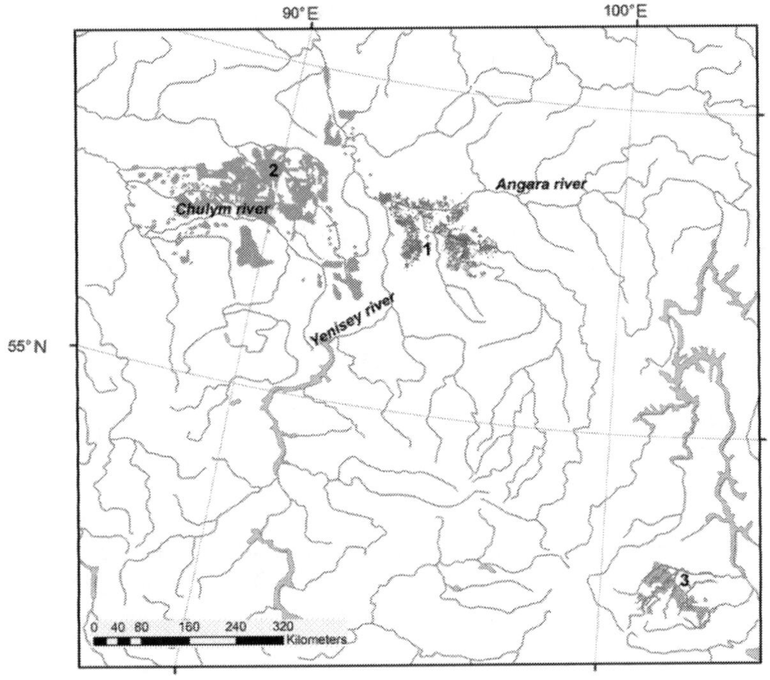

Fig. 1. Areas studied. 1 – Lower Angara area; 2 - Ket-Chulym interfluve; 3 - valley of Bolshaya Belaya river, East Sayan.

2.1 Lower Angara area

The main forest composition before silkmoth outbreak was fir, cedar, pine, larch and birch. The dead stands (mainly formed by fir) occupy a total area of 230 thousand hectares. They form several large continuous areas. Poorly damaged forests are located more mosaically.

The initial data for the analysis included:

- Landsat images acquired in 1989, 2000 and 2017;
- a map of damaged forests with three levels of damage, made in Forest Protection Center of the Krasnoyarsk Territory;
- a digital elevation model;
- a map of vegetation of the Russian Federation [5].

Using Landsat images of 1989 (before the outbreak) and 2001 (after the outbreak) and the map of dead and damaged forests with three levels of damage, the territories of the damaged and dead dark coniferous forests have been clarified. Further, these territories were analyzed using remote sensing data Landsat8 OLI of 2017 and the map [5] for each level of forest damage to identify actual land cover.

2.2 Ket-Chulym interfluve

As a result of the outbreak, which took place in 1952-1957, about 2.5 hectares of the taiga were damaged, about 1.5 million hectares of dark coniferous forests dried up, and in the territorie of dead trees in 1966, in most cases there was no satisfactory restoration of not only conifers, but also deciduous forests [1].

The initial data for these area are:

- a map of dead and damaged in 1952-1957 year forests made in Forest Protection Center of the Krasnoyarsk Territory;
- Landsat images (8 scenes for 2017, 2000 and 1978);
- a digital elevation model,
- a map of vegetation of the Russian Federation [5].

The map of dead and damaged forests was used as a mask to exclude territories outside silkmoth outbreak from the map of vegetation [5] and Landsat scenes. Landsat MSS, TM and OLI images with channels in visible, near-infrared and short-wave-infrared ranges were classified by the ISODATA method to define the land cover and enhance the map of vegetation.

2.3 Valley of Bolshaya Belaya river, East Sayan

The oldest of the considered Siberian silkmoth outbreaks occurred in the Eastern Sayan mountains in the Belaya River valley. According to [4], the beginning of large damages of cedar (Siberian pine) taiga in this region dates back to 1885. The cedar forests in the investigated area, occupied a continuous strip stretching for several tens of kilometers in the direction of the loaches, have had a continuous character at altitudes from 700 to 1700 m. In 1931, the area of damaged forests consisted of withered cedar plantings, burnt areas, with low birch and burnt areas in valleys of rivers.

For the study area the following data were used:

- a map of damaged Siberian cedar forests, compiled in 1931 (in the paper) in time of forest patology expedition [4];
- modern Landsat images of moderate spatial resolution;

- a map of vegetation cover of the Russian Federation [5];
- a digital elevation model (DEM).

The paper map of dried cedar forests was scaned and georeferenced to a Landsat image. Landsat images of 1978 (Landsat MSS) and 2013 (Landsat8 OLI) with channels in visible, near-infrared and short-wave-infrared ranges were classified to define land cover types in the territory.

3 Results and discussion

3.1 Study area in Lower Angara area

Currently in this area, only 15% remained of dark coniferous forests, mainly in areas with a low level of damage (Table 1, Fig. 2). Deciduous, pine and larch forests account for 14%. Mostly deciduous and mixed forests together are 27%; mostly coniferous – 9% of the territory. Bare soils (mainly on fire scars) and graminoid communities occupy 1.7% and 32.5% respectively where dark coniferous forests were in large areas and heavily damaged or dead. Thus, the territory of the dead forests is represented in 20 years mainly by fire scars and graminoid communities with deciduous forests that were part of the primary mixed fir forest.

Table 1. Land cover types in the areas of Siberian silkmoth outbreaks

Years of sikmoth outbreak	Dark-needle coniferous forest	Deciduous and light-needle coniferous forest	Mainly deciduous forest	Mixed forest	Mainly coniferous forest	Graminoid communuties	Bare soils
1885	58	16	5	5	2	-	1
1953-1957	7	50	7	8	18	-	
1994-1997	15	14	27		9	32.5	1.7

3.2 Study area in Ket-Chulym interfluve

In the territory of 2.5 million hectares of damaged forests, now dark coniferous forests account for 7%, pine and deciduous forests – 50%, mixed predominantly deciduous forests – 7%, mixed forests – 8%, mixed with the advantage of coniferous ones –18% (Table 1, Fig. 2). Dark coniferous forests are restored through the replacement of tree species (from deciduous birch and aspen to dark coniferous fir), and primarily in areas most closely located to undamaged dark coniferous stands. Large areas of dead woods are at the succession stage of deciduous forests. Since there was a large number of fires in the damaged forests, in these areas there is mainly deciduous forest now. The main condition for a more successful restoration of dark coniferous forests is the proximity to undamaged forests that are the source of seeds.

3.3 Study area in Belaya River valley, East Sayan

The total area of damaged forests is estimated at several hundreds of thousands of hectares, as it is written in [4] and as shows the results of paper map georeferensing and digitaizing.

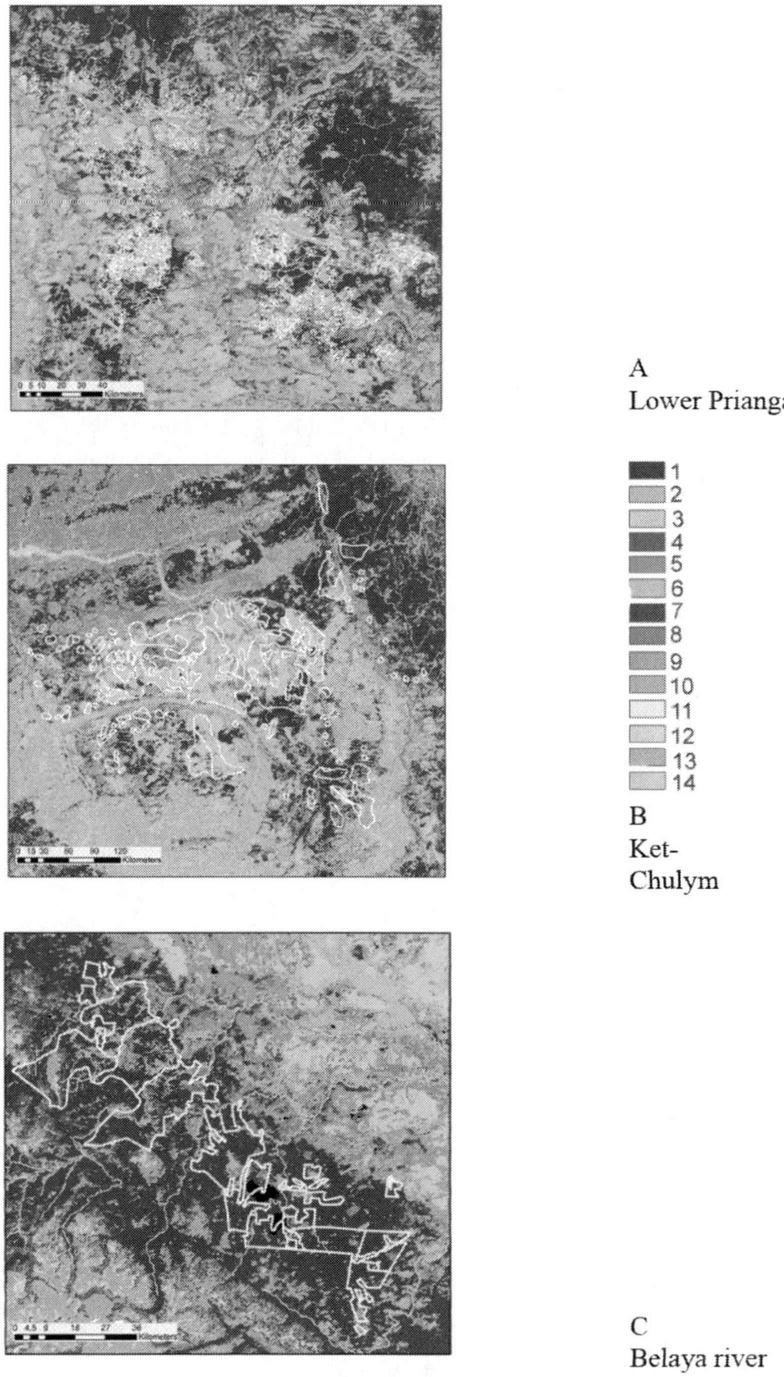

Fig. 2. Land cover types in the areas of interest: 1 - Evergreen Dark Needleleaf Forest, 2- Evergreen Light Needleleaf Forest, 3 - Deciduous Broadleaf Forest, 4 - Deciduous Needleleaf Forest, 5 - Peatlands, 6 - Humid Grasslands, 7 - Broadleaf/Needleleaf Forest, 8 - Mixed Forest, 9 - Needleleaf/Broadleaf Forest, 10 - Bare Soil and Rock, 11 – Steppe, 12 – Water bodies, 13 – urban area, 14 – Croplands

As of 1978, according to classification results of Landsat images dark coniferous forests occupied 48% of the territory, that was as dead cedar in 1931. Table 1 and Figure 2 show the percentages of different land cover types in the dry cedar forests that have been defined, using the vegetation map [5] and results of Landsat images classification. Only 58% is occupied by dark coniferous forests. The rest of the territory is occupied by pine, deciduous and mixed forests. Pine forests occupy the southern and south-western slopes, as the analysis with SRTM DEM shows.

A study of the succession process using GIS technologies can be useful for predicting the taiga restoration in the territory of the current Siberian silkmoth outbreak in the Yeniseisk district of the Krasnoyarsk Territory and adjacent territories (fig. 3). The area of dead forests is about 730 thousand hectares. Forests are damaged or dead in large areas, often limited by stands that are not food for the Siberian silkworm. Only in the east of the area there are not very large dark coniferous forests left intact, and in the west by the Yenisei Ridge.

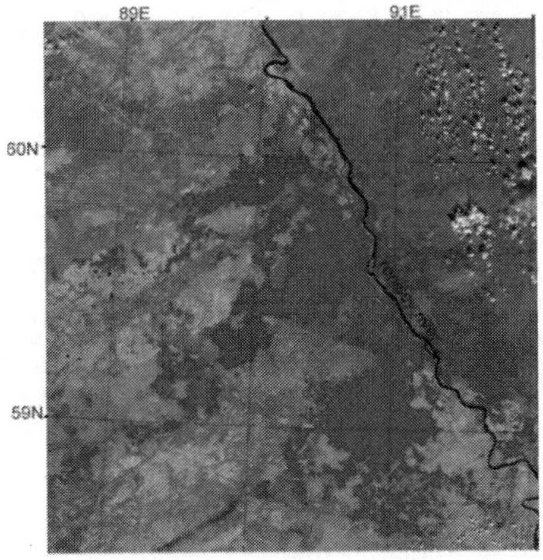

Fig. 3. Broken forests as a result of the outbreak Siberian silkworm in 2014-2017

In 20-30 years, the areas of the dead forests will be in the state of graminoid communities, and if the confluence of circumstances succeeds, one can expect that in this territory deciduous (mainly birch) forests will have grown by 2080, under the canopy of which the fir stands will have been restored.

The reported study was funded by Russian Foundation for Basic Research (No.18-47-242002), Government of Krasnoyarsk Territory, Krasnoyarsk Regional Fund of Science, to the research project: «The development of technology for creating intelligent information systems of object-oriented monitoring of territories based on remote sensing data».

References

1. V.V. Kuzmichev,V.P. Cherkashin, M.A. Korets, I.A. Michailova, Lesovedeniye, **4,** 8-14 (2001) (in Russian)
2. N.F. Reymers, Trudy po lesnomu chozyaistvu Sibiri – Works about forestry in Siberia, **4**, 255-261 (1958) (in Russian)
3. http://www.grandars.ru/shkola/geografiya/sukcessiya.html
4. V.A. Povarnitsin, *Angarskaya lesnaya expediciya 1931 – Angarsk forest expedition* (1934) (in Russian)
5. The Map of vegetation cover of Russia http://smiswww.iki.rssi.ru/default.aspx?page=317

Studying the thermal field dynamics in the forest harvesting sites

Vladimir Khamedov *

Yugra State University, Khanty-Mansiysk, 628012, Russia

Abstract. The article presents the results of the investigation of thermal field dynamics in the forest harvesting sites with the Landsat 8 satellite images in a long Infrared range. The temporary (annual and seasonal) dependences of the forest floor temperature in the harvesting sites are studied. The dependence of the forest floor temperature condition in the harvesting site on the area of the site is considered. The method to detect the sites strewn with the forest residues by using the images in a near Infrared range is described.

Keywords: space images; IR thermal image; Landsat-8; forest logging residues; the way of detection; fire danger period.

Satellite image data in infrared (IR) part of the spectrum are widely used for studying thermal contrasts of terrestrial landscapes. Special attention is given to studying "urban heat islands" [1-5]. Studying thermal regularities of the forest territories providing heat and water regulating functions is also very important. There are entire papers dedicated to studying forest temperature conditions, for example [6-7]. Forest temperature conditions are exposed to the influence of various technogenic factors, such as forest harvesting. Forest harvesting changes microclimatic conditions raising air temperature in comparison with a microclimate under a forest canopy. Temperature condition and air humidity fluctuate much stronger in the forest harvesting sites than under a forest canopy [8]. The research of temperature dependences in the forest harvesting sites will contribute to the solution of the tasks of forest rational exploitation.

The study area was located within the territory of Khanty-Mansi Autonomous Okrug, which has considerable forest resources. Forests cover an area of about 50,41 million hectares (6,2% of the total forest area in Russia) with the planting stock of 3133,14 million m^3. More than 130 companies and entrepreneurs are involved in logging business. By January 1, 2018 there have been signed 108 forest logging agreements. The total annual amount of harvesting wood on lease contracts is 2876,21 thousand m^3, and the area of forests transferred to leasers is about 2812,085 thousand hectares.

This paper examines the features of thermal field dynamics for the forest harvesting sites obtained with the Landsat 8 spacecraft in a long Infrared range and also the method of detecting the sites strewn with the forest residues by using the images in a near Infrared range. The site of 113800 hectares located in a southwest part of the district has been chosen as a study area.

* Corresponding author: khamedov.vladimir@gmail.com

Long range Infrared imaging

The thermal field dynamics was investigated in 22 forest harvesting sites of different area located within the borders of the test site. The minimum size of harvesting sites was 2,27 hectares, the maximum – 26,97 hectares. The research was conducted for the time period from 2014 to 2017.

Sets of space images used for the research are presented in table 1.

Table 1. Frames from orbiting satellites made in different periods

Year of imaging	Frame	Time of imaging, hour. (UTM)
2014	LC08_L1TP_162017_20140703_20170421_01_T1	06:48
2015	LC08_L1TP_162018_20150722_20170406_01_T1	06:47
2016	LC08_L1TP_162018_20160606_20170324_01_T1	06:47
2017	LC08_L1TP_161018_20170618_20170629_01_T1	06:42

Terrestrial radiation or thermal infrared radiation is determined by the temperature and composition of the radiating objects displayed on the satellite image. The surface radiation temperature for each image is reconstructed based on the spectral radiation intensity which was recorded by the imaging device [9].

The calculated mean instantaneous temperatures within and beyond (background value) the borders of forest harvesting sites by years are presented in table 2. The background value was calculated as the arithmetic mean of the frame temperature, excluding cloudy areas and forest harvesting sites. The value of air temperature at an altitude of 2 meters above the Earth's surface was available from the nearby weather station in the village of Leushi (WMO ID 28064), located 23 km to the center of the test site. For all years, the temperature values recorded by the weather station at 11:00 local time were used

Table 2. The measured values of the surface temperatures in different years

Date of measurements	Measured maximum (peak) value of temperature on the forest harvesting site, °C	Measured mean temperature, °C		Temperature difference Δtav, °C	Air temperature available from the nearby weather station, °C
		Forest harvesting sites	Background value		
03.07.2014	30,31	24,36	19,22	5,15	20,7
22.07.2015	26,47	22,91	20,57	2,34	23,2
06.06.2016	26,74	21,36	18,87	2,44	19,4
18.06.2017	26,11	22,39	19,87	2,52	22,2

The calculated value of temperature difference between the mean temperature within the borders of the forest harvesting site and the background value of the temperature is shown in figure 1.

In the first year of measurements the mean temperature difference between the forest harvesting site and surrounding background was 5,15 °C. The temperatures leveled off in the next years with the mean difference 2,43 °C. The total absence or partial forest litter destruction, open drylands, logging residues on the forest harvesting sites are the most likely causes for the anomaly in the first year. Then, in the course of the constructive process, temperatures on the forest harvesting sites leveled off.

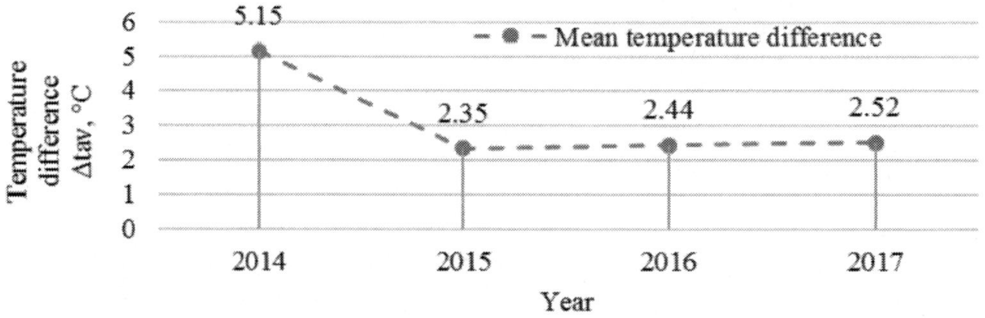

Fig. 1. Dynamic analysis of the mean temperature difference between the forest harvesting site and the background value

The dependence of maximum (peak) surface temperatures within the borders of the forest harvesting site on the area is presented in figure 2.

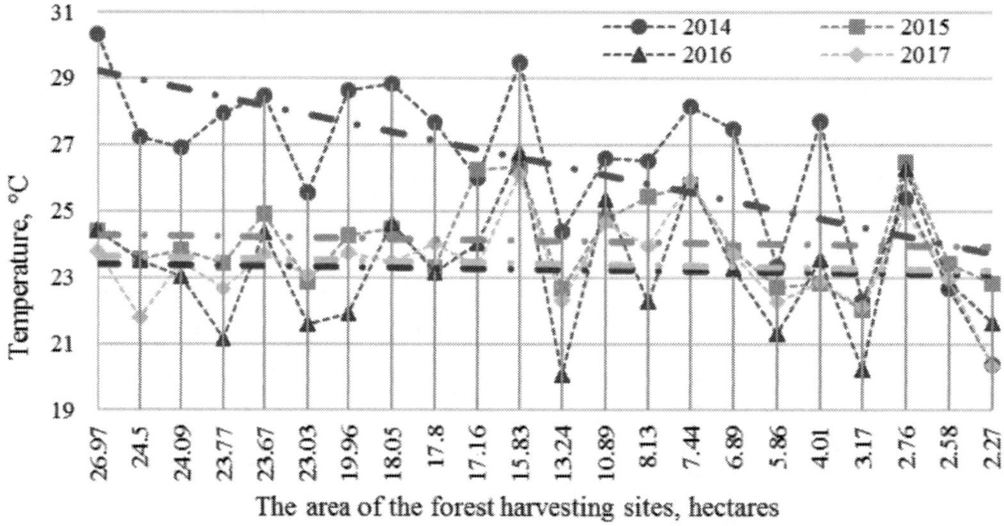

Fig. 2. Dependence of surface temperature in forest harvesting sites on the area

In the first year there was marked positive correlation between the surface temperature in the forest harvesting site and the area of the site (the red line on the diagram). In the next years this dependence remained, however it was not so significant.

Seasonal dynamics of surface temperature in the forest harvesting sites was examined on the example of 71 sites in the period of time from January to December, 2017. The results of studying seasonal dependence of temperature difference are presented in table 3, and graphically in figure 3.

Near range Infrared imaging

Some amount of biomass is often left on sites after harvesting operations (i.e. branches, tops, stumps). According to Fire safety rules it is prohibited to leave logging residues on the harvesting sites during the fire danger period [10]. Now space imagery of high spatial resolution or aerial photographs are used for detecting the sites with the logging residues. The logging residues are visually interpreted on such photographs. However, images of high spatial resolution have, as a rule, a limited swath width and cost a lot.

Table 3. Mean temperature within the borders of forest harvesting sites and its background value in 2017 , t°C.

Date of measurements	09.01.2017	10.02.2017	14.03.2017	21.03.2017	22.04.2017	01.05.2017	25.06.2017	15.10.2017	25.11.2017	18.12.2017
Mean temperature within the borders of the forest harvesting site, °C	-34,71	-27,36	-4,82	-1,98	6,61	9,51	18,19	3,35	-12,26	-13,78
Mean background temperature °C	-35,26	-27,96	-5,29	-2,77	-5,46	7,86	14,61	3,13	-12,57	-13,64
Temperature difference Δt, °C	0,54	0,60	0,47	0,79	1,15	1,65	3,58	0,22	0,32	-0,14

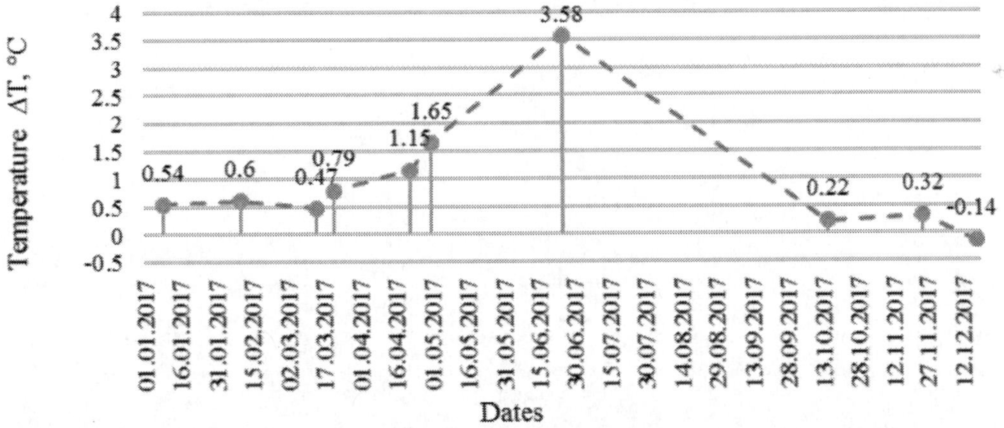

Fig. 3. Seasonal dependence of the mean temperature difference in the forest harvesting sites

During the spring thaw period the sites with the logging residues can be found by multispectral space images in the Infrared range [11], according to the Kirchhoff's radiation law, due to different surface temperatures on the sites. When snow cover melts in this time of the year the logging residues are seen on space images as thermal anomalies in the near infrared range. The difference in brightness between the pats of the image is caused by various albedo of the warmed sites.

A multispectral image for interpreting the sites with the logging residues can be obtained with the OLI instrument of Landsat 8, RGB combination (7-5-3).

In figure 4 fragments of space images produced by Landsat-8 at different times in the Infrared range on the forest harvesting site near the settlement of Leushi, Kondinsk Region, Khanty-Mansi Autonomous Okrug are presented. A collection of images allows to monitor the dynamics of forest harvesting sites appearance and the character of their thermal radiation. The total area of the forest harvesting site represented on the top part of the image is 13,27 hectares. The area of the site was formed in the following order: the northeast part with an area of around 4,3 hectares was cut down in a time interval between January 9, 2017 and January 25, 2017 (LC81610182017025LGN00). The northwest (the area of 4,81 hectares) and the southeast part (1,94 hectares) – in the period of time between January 25, 2017 and February 10, 2017 (LC81610182017041LGN00). The southwest part with an area of 2,22 hectares – in the period of time between February 10, 2017 and March 14, 2017 (LC81610182017073LGN00).

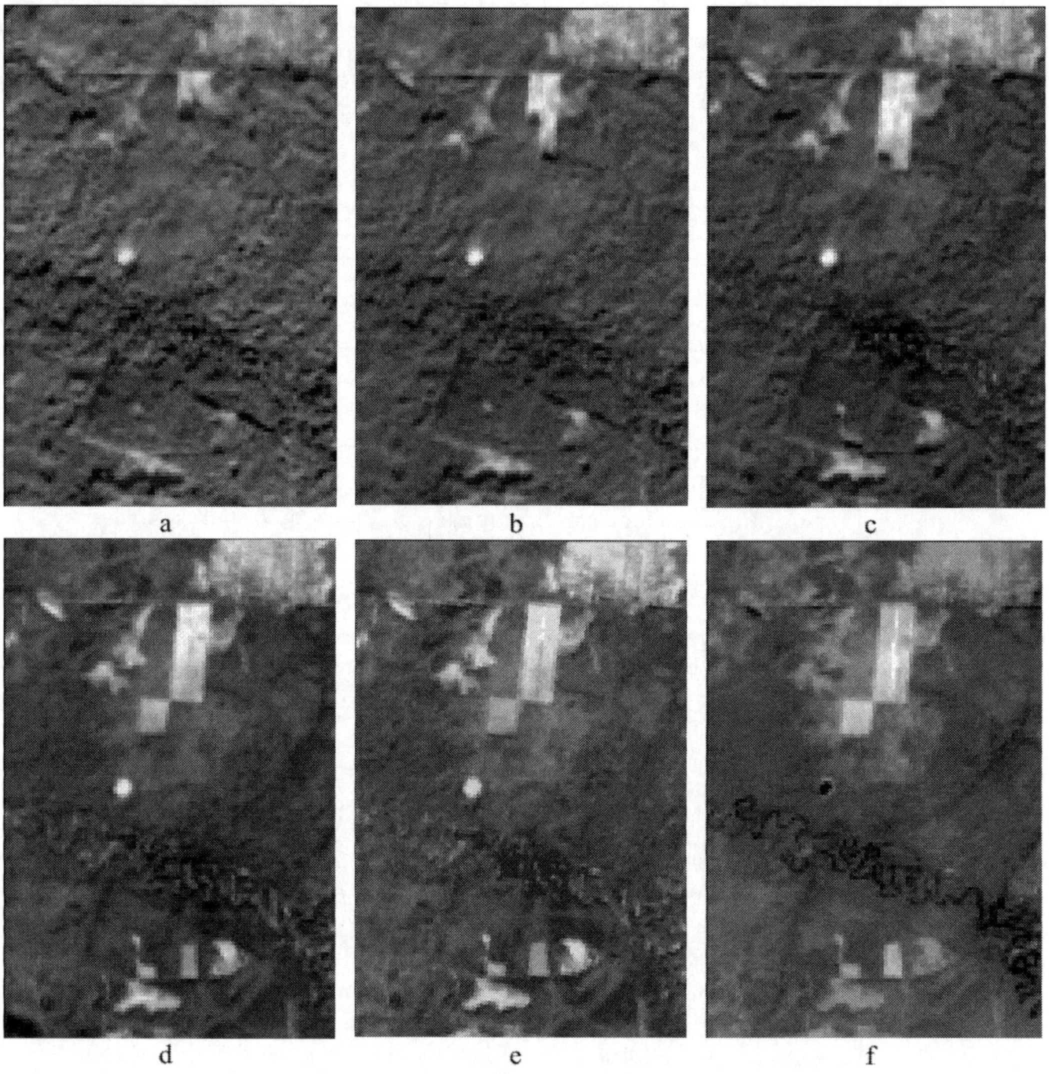

Fig. 4. Images of forest harvesting sites with logging residues produced in the Infrared range at different times (combination of spectral bands 7-5-3): a – January 9, 2017; b – January 25, 2017; c – February 10, 2017; d – March 14, 2017; e – March 21, 2017; f – May 1, 2017

The site with an area of 6,83 hectares adjoining from the South was cut down between February 10, 2017 and March 14, 2017. The site to the south with an area of 3,86 hectares was cut down between February 10, 2017 and March 14, 2017.

Consider the image of forest harvesting sites shown in figure 4. The image was obtained on March 21, 2017 during the initial stage of snow melting. Objects with thermal anomalies – the sites with the logging residues are depicted with red. On the site of 13,27 hectares the logging residues collected into piles are seen.

Figure 5 shows the combined diagrams of the current snow depth and the air temperature available from the nearby weather station in the village of Leushi (source: http://rp5.ru). For this territory positive dynamics of snow melting was observed from February 21, 2017. With an increase in temperature, sites with the logging residues warmed up for a shorter period of time. The site warmed up as temperature increased and snow cover melted. From April 6, 2017 the depth of snow cover decreased sharply, complicating detection of sites with the logging residues (figure 4-f).

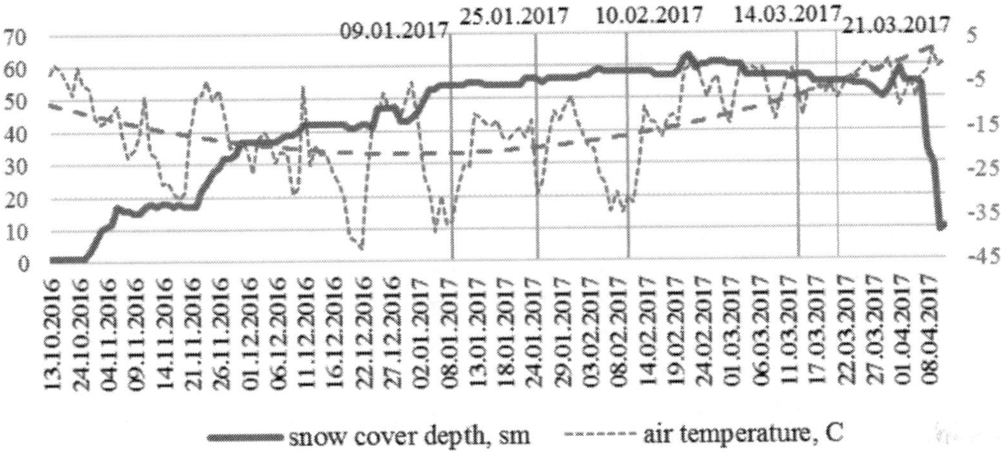

Fig. 5. Dynamics of snow cover depth and air temperature in the observation area

The method presented can be used to detect sites with the logging residues in the process of interpreting space images with an average spatial resolution in the period of intensive spring snow melting. This method allows to find the forest harvesting sites with the logging residues left in the previous winter period, define the borders and the areas of those sites. The data collected can be used for carrying out complex satellite monitoring of forest ecosystems [12, 13] and for operational fire danger forecasting in the observation areas just before the beginning of the fire danger period.

Conclusion

Studying the thermal field dynamics in the forest harvesting sites helps to ensure rational exploitation of forest resources by maintaining the proper ratio between the amount of forest exploitation and the conservation-oriented policies. The method suggested to detect the forest harvesting sites with the logging residues is based on the use of space images with an average spatial resolution. The use of Infrared for detecting the sites with the logging residues has seasonal character. Therefore the method suggested can be applied in the period of intensive snow cover melting. The possibility of detecting and mapping the sites with the logging residues just before the beginning of a fire danger season can be referred to the practical importance of the research. This will allow to take the measures necessary for decreasing fire danger in the forest territories before the fire danger season.

References

1. Oki K., Omasa K. Technique for Mapping Thermal Infrared Radiation Variation Within Land Cover // IEEE Transactions on Geoscience and Remote Sensing. 2003. Vol. 41. № 6. P. 1521-1524.
2. Gluch R., Quattrochi D.A., Luvall J.C. A multi-scale approach to urban thermal analysis //Remote Sensing of Enviromment. 2006. № 104. P. 123-132.
3. Ilyin Yu.A., Chufarova N.E. Complex analysis of thermal infrared imagery of the earth surface//News of higher educational institutions. Geodesy and aerial photography. – №. 2. – 2011 – P. 37 – 40.
4. Preston N.E. Features of spatial distribution of radiation temperature for different landscapes on thermal infrared images//News of higher educational institutions. Geodesy and aerial photography. – № . 5. – 2012 – Page 63 – 66.
5. Gryshchenko M.Yu. Application of thermal ETM+ imagery for studying Moscow urban heat island /Modern problems of remote sensing of Earth from space. 2012. V . 9. № . 4. Page 95-101.
6. Sakharov M.I. Dependence of soil temperature on the character of forest cover//"Soil science". 1948. № 3. - Page 5-13.
7. Arkhipova E.P. The Effect of of forest belts on the soil temperature profile//Works of the main geophysical observatory of A.I. Voyeykov. 1952. Issue 36 (98). Page 27-38.
8. Kozlov V.M. The Effect of Logging on Habitat and Populations of Game Animals of the European Taiga monograph. – Kirov: Vyatka GSHA, 2010. - 150 p.
9. Chen Y., Sui D. Z., Fung T., Dou W."Fractal analysis of the structure and dynamics of a satellite-detected urban heat island"//International Journal of Remote Sensing, 28, № 10, 20 May 2007, 2359–2366.
10. The resolution of the Government of the Russian Federation dated June 30, 2007 №417 "Fire safety rules" Access mode: The GARANT Legal Information Resource
11. Khamedov V. A. Detection of forest sites with logging residues by medium spatial resolution satellite imagery//Materials of the III All-Russian scientific and practical conference "Geoinformation Technologies for the Solution of problems of Rational Environmental Managemen" Khanty-Mansiysk. 2017, 72-75.
12. Khamedov V. A., Mazurov B.T. Development of methodical issues for the creation of forest ecosystems satellite monitoring system in conditions of oil and gas complex influence in the territory of Western Siberia//Vestnik SGUGiT . 2015. Issue 3 (31), 16–31.
13. Mazurov B.T., Avrunev E.I., Khamedov V.A.. Operational monitoring of forests in northern regions using optical and radar space imagery//Modern problems of remote sensing of the Earth from space 2017. V. 14. No. 4, 103-111.

Analysis of GRACE satellite measurements over seismically active areas of the strongest earthquakes

Konstantin V. Simonov[1,*], *Valentin* B. Kashkin[2], *Tatyana* V. Rubleva[2], and *Konstantin* V. Krasnoshekov[1]

[1]Institute of Computational Modeling SB RAS, Krasnoyarsk, 660036, Akademgorodok, Russian Federation
[2]The Institute of Engineering Physics and Radioelectronics of Siberian Federal University, Krasnoyarsk, 660074, Akademika Kirenskogo Str. 28/12B, Russian Federation

Abstract. The study is devoted to the analysis of the features of the change in the parameter EWH (Equivalent Water Height) over the geoid from the satellite measurements of the GRACE space system in seismically active areas of the strongest underwater earthquakes. The GRACE satellite data archive was created. An interpretation of the disturbed state of the geomedium using digital maps of the spatial distribution of the parameter EWH is performed.

1 Introduction

Methods for studying the gravitational field from satellite measurements were developed more than 40 years ago. However, only in recent years in modern space gravity missions such as CHAMP (Challenging Mini-Satellite Payload, DLR, 2000-2010), GRACE (Gravity Recovery and Climate Experiment, NASA / DLR, 2002-2017) and GOCE (Gravity Field and Steady-State Ocean Circulation Explore, ESA, 2009-2013) continuous GPS tracking for low-orbit satellites has been implemented [1].

With the help of the CHAMP, GRACE, and GOCE systems data on the Earth's gravitational field were obtained with high accuracy and high spatial resolution. Thus, the accelerometer of the CHAMP satellite had an accuracy of $3*10^{-9}$ m/s^2, which is an order of magnitude smaller than the accuracy of the GRACE device equal to $1*10^{-10}$ m/s^2 [2]. The archive of data obtained with the help of GOCE satellite equipment contains data for only 4 years, while the GRACE information archive contains satellite measurements for 15 years, including data on space-time variations of water reserves.

Note that in the GRACE space project satellite-satellite technology is implemented using the low-low scheme, the essence of which is to measure distances and rates of change in the distance between them. Two identical GRACE satellites move 170-240 km one after another along a circular circumpolar orbit with an inclination of 89° [1]. The altitude of this

* Corresponding author: simonovkv@icm.krasn.ru

orbit is about 450 km. The orbital plane of each satellite is affected by disturbing accelerations, which correspond to the first derivatives of the gravitational potential.

The equipment of GRACE satellites consists of: a highly stable frequency generator, a K-band Ranging system, a GPS receiver, an astroblock, corner reflectors for laser location, 2 accelerometers ACC and SuperSTAR. The accuracy of measuring the distance between GRACE satellites is determined by rangefinders up to 10 μm, which allows obtaining data on variations in gravitational force on the basis of differential measurement methods in systems with variable geometry of the arrangement of elements [3]. Satellites fly over a certain section of the Earth once every 30 days, providing a complete mapping of the earth's surface.

In this paper we studied the features of the variations of the equivalent water level (parameter EWH) over the geoid contour from the satellite measurements of the GRACE space system in the seismically active regions of the strongest earthquakes in order to reveal the connection between the stressed-deformed state of the geomedium and the processes of preparing the expected main seismic event.

The strongest underwater earthquakes with magnitudes [4] were chosen as objects of research: M = 9.1 (December 26, 2004, near Sumatra Island); M = 8.1 and M = 8.3 (November 15, 2006 and January 13, 2007, in the Simushir region); M = 8.8 (January 27, 2010, near the coast of Chile), M = 9.1 (March 11, 2011, near the east coast of Honshu Island) and M = 8.3 (on May 24, 2013, near the Kamchatka Peninsula in Sea of Okhotsk).

2 The gravitational field and the parameter EWH

Satellite measurements of GRACE COP of the water surface height with respect to the geoid contour allow determining the deviation of the free surface of the sea from its average level along the satellite path. A geoid is an equipotential gravity field, which corresponds to the mean sea level at rest [5]. Monthly data obtained from the site [6] contain information on the deviation of the surface of the World Ocean from the geoid in units of the equivalent water level EWH (in cm).

In general, the parameter EWH is used to analyze the dynamic topography of the oceans and refine the parameters of the Earth's external gravitational field [7]. The method of decomposition of the Earth's gravitational potential with respect to spherical functions is described in [5, 3]. This method of representing potential is used in the study of the gravitational field by the disturbed motion of satellites. The redistribution of water in the hydrosphere takes place in a thin layer near the surface of the Earth, just a few kilometers wide. The masses between the geoid and the water surface have a constant density $\rho_w$of 1000 kg/m^3 [8].

The EWH is calculated on the basis of the harmonic coefficients of geopotential models minus the coefficients of the middle model (EIGEN-6C) [7]. This is a global model, the initial data of which are ground gravimetric measurements, laser trajectory observations of LAGEOS satellites (1985-2010), as well as satellite information of GOCE and GRACE. In the high-power EIGEN-6C model, the geopotential decomposition occurs to a degree of the order of 1420, which corresponds to a 14-km spatial resolution [9].

Equivalent water level (parameter EWH) is found from the ratio of surface density $\Delta\sigma$ to water density ρ_w:

$$EWH = \Delta\sigma/\rho_w$$

3 Results analysis

Based on the GRACE satellite data [6], digital maps and three-dimensional images of the space-time distribution of the parameter EWH with a period of 30 days are constructed. An example of the map obtained in the isolines for December 2004 for the Sumatra earthquake region is shown in Figure 1, *a*. Figure 1, *b* shows an image of the focal region of this earthquake in the form of a 3D model for February 2005. Figure 1, *a* and Figure 1, *b* the scale of the parameter EWH (in cm) is given. The images shown in Figure 1, obtained with the help of satellite altimetry, reflect the features of the structure of the underlying geo-environment during the catastrophic event.

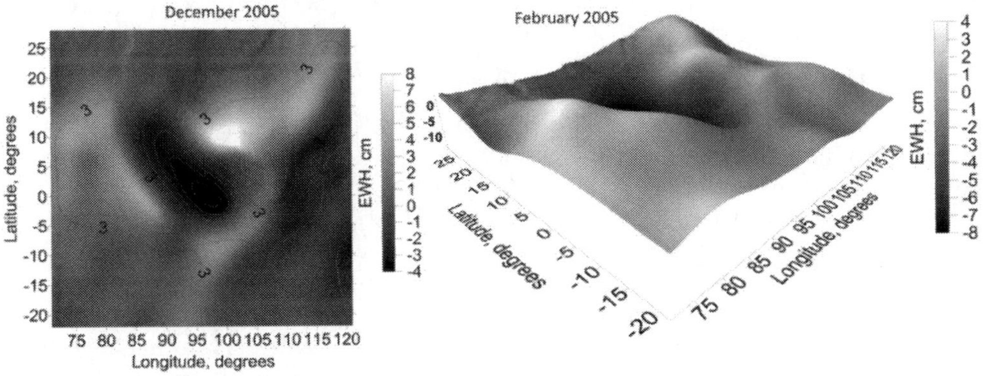

a *b*

Fig. 1. Digital map in isolines for December 2004 (*a*) and 3D models for February 2005 (*b*) distribution of the parameter EWH of the Sumatra earthquake.

Figure 2 shows a digital map in the isolines of the spatial distribution of EWH for the seismically active region of Chile in February 2010 during the preparation and occurrence of the expected strongest earthquake according to GRACE satellite data. The point in Figure 2 indicates the hypocentre of the earthquake. You can see an abnormal area with increased values of the parameter EWH. It is revealed that the hypocenter of the earthquake is located near the anomaly axis, which is oriented submeridionally, at the maximum gradients of the parameter EWH. Interpretation of the direction of the revealed anomaly is related to the geometry of the subduction zone of lithospheric plates.

 https://doi.org/10.1051/e3sconf/20197502007

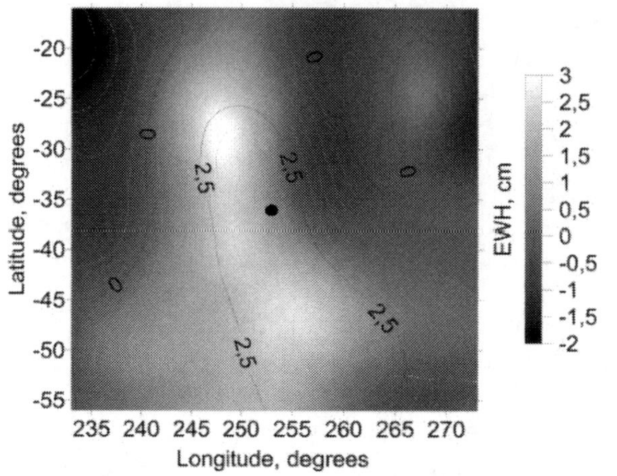

Fig. 2. Digital map in isolines of the Chilean earthquake for February 2010.

A graph of the change in EWH at the hypocentral point was plotted 5 months before the seismic event and 5 months after the earthquake Mw = 8.8 (Figure 3). The characteristic change in the parameters EWH, shown in Figure 3, shows the process of preparing the strongest earthquake and its aftershock discharge. For "0" on the horizontal time scale in Figure 3 the date of the considered seismic event is adopted. The "negative part" of this scale characterizes the process of preparing the strongest earthquake, the "positive part" is aftershock activity. The vertical scale in Figure 3 reflects the changes in the parameter EWH, characterizing the sharp increase in the water surface boundary during the process of making. Dedication is understood as the decompression of the upper layers of the geomedia and their saturation with water.

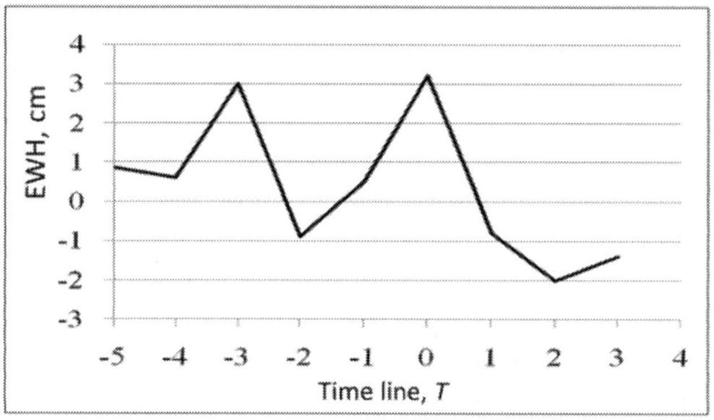

Fig. 3. The dynamics of the change in the parameter EWH with a period of 1 *T* = 30 days during the process of preparing the strongest earthquake and its aftershock discharge.

With a deep earthquake (Sea of Okhotsk, May 2013), a zone of increased EWH values was observed in the area of preparation (Figure 4). The hypocenter of the earthquake was located at a depth of 602 km and Figure 4 it is indicated by a dot. The digital map shows that the anomalous region with increased values of the parameter EWH is oriented in the SW-NE direction. From Figure 4 it becomes clear that the hypocenter of the earthquake is located near the axis of the anomaly at the maximum gradients of the parameter EWH. It is important to note that, in contrast to subduction earthquakes, which have a pronounced linear geostructure, a deep event is characterized by a localized focal zone.

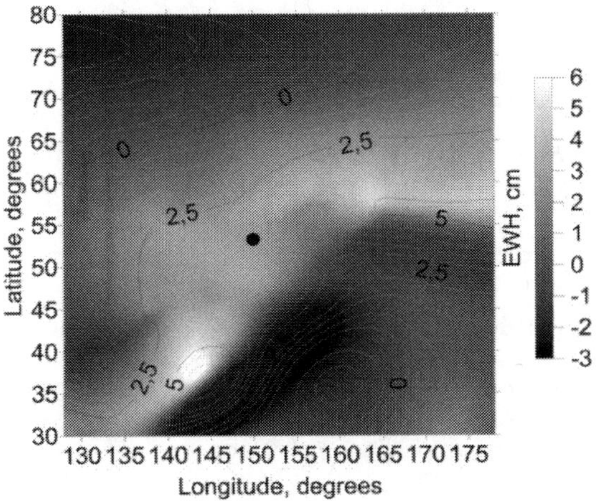

Fig. 4. Digital map in isolines of the Okhotsk earthquake for May 2013.

4 Conclusion

As a result of the research, a GRACE satellite measurement archive has been compiled containing information parameter data on the EWH (equivalent water level). The parameter EWH during the preparation and relaxation of the strongest earthquakes varies over the contour of the geoid. GRACE data have been analyzed for a number of areas of focal zones of known underwater earthquakes (2004-2013). Digital maps of the EWH variation over the geoid in the focal zones of the strongest earthquakes studied are constructed. The specified parameter can serve as a basis for an estimation of geodynamic risks of occurrence of destructive earthquakes for coastal seismically dangerous territories.

References

1. F. Flechtner, N. Sneeuw, Wolf-Dieter Schuh (Eds.), Science Report. Advanced Technologies in Earth Sciences, Springer, **20** (2014)
2. I.N. Gansvind, Electronic Libraries, **20**, 1 (2017)
3. R.A. Kashcheev, Modern methods of satellite gravimetry (Kazan. Univ., 2015)
4. USGS, Latest Earthguakes, https://earthquake.usgs.gov/earthquakes/map
5. B. Hoffmann-Vellenhof, G. Moritz, Physical geodesy (Moscow, MIIGAiK, 2007)
6. NASA, ftp PO.DAAC, ftp://podaac-ftp.jpl.nasa.gov /allData/tellus/L3/ocean_mass /RL05/ascii/
7. B. Wouters, J.A. Bonin, D.P. Chambers, R.E.M. Riva, I. Sasgen, J. Wahr, Review Article Rep., Prog. Phys. (2014)
8. J. Wahr, M. Molenaar, F. Bryan, J. Geoph. Res., Solid Earth, **103**, B 12 (1998)
9. V.F. Kanushin, A.P. Karpik, I.G. Ganagina, D.N. Golodobin, A.M. Kosareva, N.S. Kosarev, Investigation of modern global models of the Earth's gravitational field, (Novosibirsk, SSUGiT, 2015)

E3S Web of Conferences **75**, 02008 (2019) https://doi.org/10.1051/e3sconf/20197502008
RPERS 2018

ANTARCTIC OZONE HOLE AS A NATURAL GEOPHYSICAL OBJECT

Alexander V. Dergunov[1], *Valentin* B. Kashkin[2], *Tatyana* V. Rubleva[2,*], *Alexey* A. Romanov[3] and *Roman* V. Odintsov[2]

[1]Institute of Computational Modeling SB RAS, Krasnoyarsk, 660036, Akademgorodok, Russian Federation
[2]The Institute of Engineering Physics and Radioelectronics of Siberian Federal University, Krasnoyarsk, 660074, Akademika Kirenskogo Str. 28/12B, Russian Federation
[3]The Institute of Space and Information Technology of Siberian Federal University, Krasnoyarsk, 660074, Akademika Kirenskogo Str. 26/1, Russian Federation

Abstract. Satellite data on total ozone content for 1985-2015 have been used. Methods of evaluating ozone deficit in the polar region and its excess in middle latitudes of the Southern Hemisphere have been developed. In early spring the ozone molecules outflow and the ozone anomaly forms. Ozone inflows the middle latitudes, its total content increases and a ring with elevated TO forms. In October-November the dynamic process reverses, from the ring the ozone molecules transfer to the polar latitudes. The amount of ozone leaving the ring into the polar regions and filling the ozone anomaly is virtually the same. The results produces indicate that the Antarctic ozone hole is a natural geophysical formation.

1 Introduction

The ozone layer protects our planet from the hard ultraviolet rays of the Sun. Total ozone (TO) – is the thickness of gaseous ozone layer in the vertical column of atmosphere at temperature 0°C and atmospheric pressure 1013 hPa. TO is measured in Dobson units (DU), 1 DU = 10^{-5} m. The ozone anomalies are local regions with substantially reduced total ozone in the Earth's ozone layer below 220 DU.

In 1985 the Southern Hemisphere around the Antarctic in spring was found to form an ozone anomaly (Antarctic ozone hole, AOH) [1]. In recent years its area varies from 18 to 30 million km^2 [2]. AOH is surrounded by an ring with high TO, its values reach up to 450 DU. In 1986 researchers R.S. Stolarski and M.R. Schoeberl showed that during AOH formation the ozone mass is transferred from the near-polar regions of the Southern hemisphere to the middle latitudes, and from the end of September the ozone mass moves towards the pole. In November-December AOH disappears [3].

* Corresponding author: tvrubleva@mail.ru

However, currently is generally accepted the theory holding that the cause of ozone layer depletion in the polar regions of the Southern hemisphere is destruction of ozone by some substances of anthropogenic origin for example chlorfluorcarbons (freons) [4, 5]. Vienna Convention (1985) and the Montreal Protocol (1987) on Substances That Deplete the Ozone Layer were signed. When signing the Montreal Protocol it was claimed that the Antarctic ozone hole would disappear in 2010, however, this anomaly occurs annually.

The work studies redistribution of ozone from the middle latitudes to the polar regions of the Southern Hemisphere in spring 1985-2015 by TOMS (Nimbus-7, EarthProbe, USA) and OMI (Aura, USA) satellite data [6]. Total ozone was measured in the range of 300-380 nm with an accuracy of 2-4%.

2 The ozone spring dynamics

Mean zonal TO was analyzed from NASA site [6]. Mean zonal TO are daily values of total ozone in the different 5 degrees rings of latitudes averaged by longitude from 0 to 360 degrees.

To illustrate the dynamic processes running in spring 2013 let us consider plots of mean zonal TO vs. individual days in October and November given in Figure 1. Figure 1,*a* shows variation of ozone content during the existence of the Antarctic ozone hole. It is evident that over the polar region in October the mean zonal TO values are below the climate normal 220 DU. E.g. on the 1st of October in the ring of latitudes 85-90°S the zonal TO is 142 DU. In the middle latitudes the total ozone layer substantially exceeds the climate normal and on the 11th of October in the latitude ring 50-55° S TO is 353 DU. So in the range of latitudes 45-55°S a ring with high the total ozone forms.

Dynamic processes during the existence of AOH have been considered in [7]. Account is taken of the enormous circumpolar vortex in the Southern Hemisphere reaching over the polar and middle latitudes. In August-November in the stratosphere this vortex is the high-pressure area with center in the polar region. Ozone flows (together with the main mass of air) move from the pole in all directions, curving by the Coriolis force from west to east. This forms rotating vortex at the middle latitudes. The vortex absorbs ozone from the polar latitudes and includes the ozone ring with high TO. Ozone concentration within the vortex drastically drops and an area with reduced ozone content including AOH forms over the pole. By the end of spring the Antarctic stratosphere is heated, the vortex grows into a cyclone, the ozone mass reverses from the ring to the polar region, the anomaly disappears.

Lack of satellite data on ozone content in the polar regions in August-September denies the opportunity to study the ozone mass transfer from the polar area into the middle latitudes. However, these data available for October-November make possible to compare the amount of ozone that has left the polar latitudes and ozone excess in the middle latitudes.

Figure 1,*b* shows variation of zonal TO during active atmospheric processes with AOH filling in November. We should note that in the polar and middle latitudes these values exceed the climate normal. In the polar regions the ozone content increased, in the middle latitudes it decreased. By the end of November the high TO ring in the middle latitudes disappears (29 November, Figure 1,*b*), and the ozone hole ceases to exist.

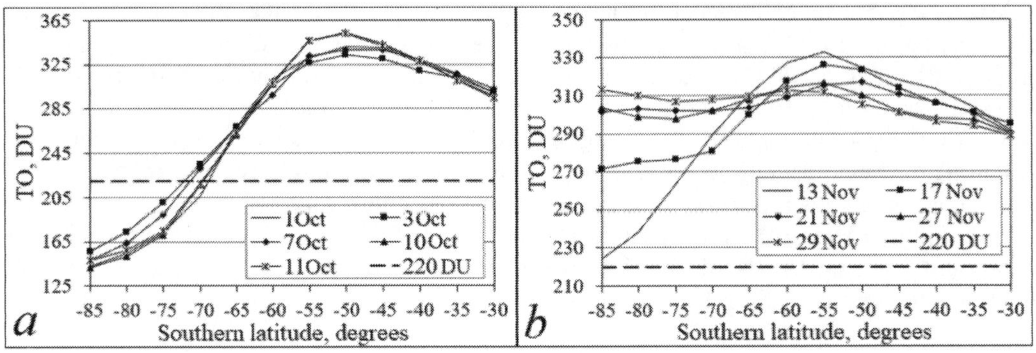

Fig. 1. Variation of ozone content in the Southern Hemisphere in spring 2013.

3 Select averaging interval

To improve validity of the analysis we studies a plot averaged zonal TO of vs. latitude over several days. It was necessary to define time averaging interval ΔT.

From the moment of its formation to «filling» the Antarctic ozone hole is known [7] to exist for 3-4 months, i.e. actually from the 20-s of August to mid-December. Zonal data on the total ozone from August till the end of September are not available [6]. During this period the lower stratosphere above the Antarctic is not illuminated by the sun and measurements were not performed. For the beginning of the time interval ΔT we chose the data for which the NASA base showed the TO data. In 2013 it's the 1st of October.

To define the end of interval ΔT when the ozone hole «filling» processes start we analyzed deficit of ozone mass ΔM. According to [2] this characteristic is defined as the difference between the ozone mass in AOH with the boundary of 220 DU, and the mass of ozone within actual boundaries of the anomaly. Comparison of zonal TO and ΔM deficit of AOH showed that the data of beginning of drastic restoration of the ozone layer in the polar regions in 2013 is the 14th of October. This date was chosen as the end of time interval ΔT. So, in 2013 ΔT = 14 days.

Figure 2 shows variation of zonal TO in the Southern Hemisphere in two latitudinal zones 75-90° S and 45-60° S from the 1st to the 22nd of October 2013. The dash lines delineate boundaries of the earlier chosen ΔT interval. Figure 2,*a* indicates income of ozone into the AOH area. Growth of zonal after the 14th October confirms correct choice of the upper boundary of the interval ΔT.

From the 10th of October the zones of 45-60°S lying beyond the polar regions are observed to decrease zonal TO. One can talk about ozone transfer from the middle latitudes into the Antarctic ozone hole. TO is observed to drastically decrease after the 14th of October. So, the data presented in Figure 2 do not contradict the method of choosing the upper boundary of the interval ΔT in ozone deficit value ΔM in AOH.

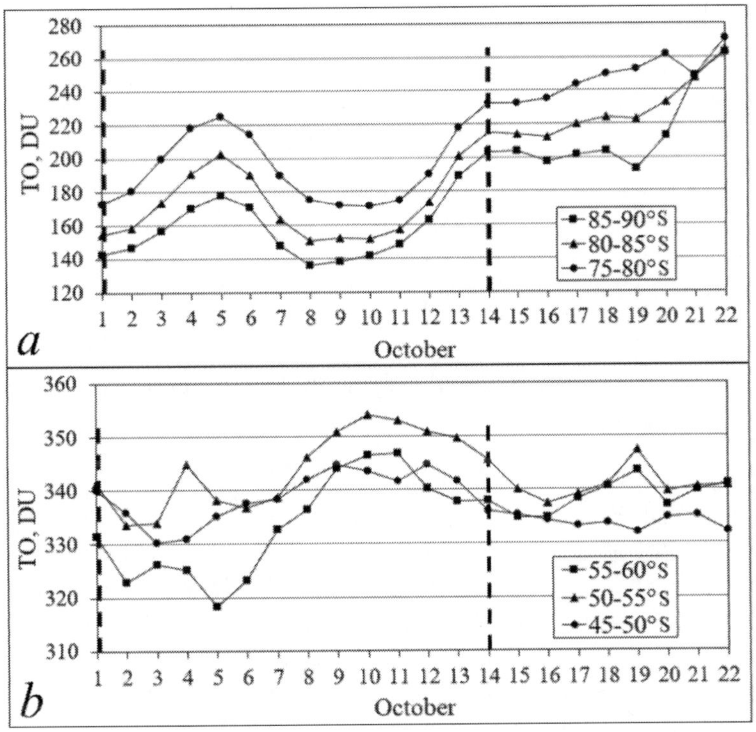

Fig. 2. Variation of zonal TO over interval ΔT (boundaries of the interval ΔT are indicated by the dash lines) for the polar latitudes *(a)* and for the middle latitudes *(b)*.

4 Results analysis

Let us compare the amount of ozone that has «left» the polar regions and ozone excess in the middle latitudes. For this we need to define ozone excess in the middle latitudes and its deficit in the polar regions. For this we choose some threshold value of TO, denote it as X_0. Exceeding the X_0 values correspond to the excess, and below the X_0 threshold we observe deficit. The threshold value of X_0 it was experimentally found to be the annual mean total ozone content in the range of latitudes from 32.5° to 52.5°S for every year. According to the NASA database all zonal mean are present in this latitude range without omissions. For 2013 the threshold value of $X_0 = 295.5$ DU.

Figure 3 shows deviation of ozone content in the polar and middle latitudes of the Southern Hemisphere from X_0 threshold in the chosen interval ΔT. In Figures 3,*a* and 3,*b* the threshold value of X_0 is indicated by the dash line. The plot presented in Figure 3,*a* matches the plot in Figure 1,*a* with the only difference that the plot in Figure 1,*a* represents dependence of daily zonal means vs. geographic latitude φ, while the plot in Figure 3,*a* – daily zonal means are averaged over 1 – 14 October. Denote the zonal values averaged in this way by x_φ. In Figure 3,*a* it is evident, that the amount of ozone x_φ in the polar region is less than the threshold X_0 at the latitudes above 63°S. This area might be specified by ozone deficit, and in Figure 3,*a* it is denoted by «–». In Figure 3,*a* at the latitudes below 63°S the amount of ozone x_φ exceeds the threshold TO. This area is specified by ozone excess, it is denoted by «+».

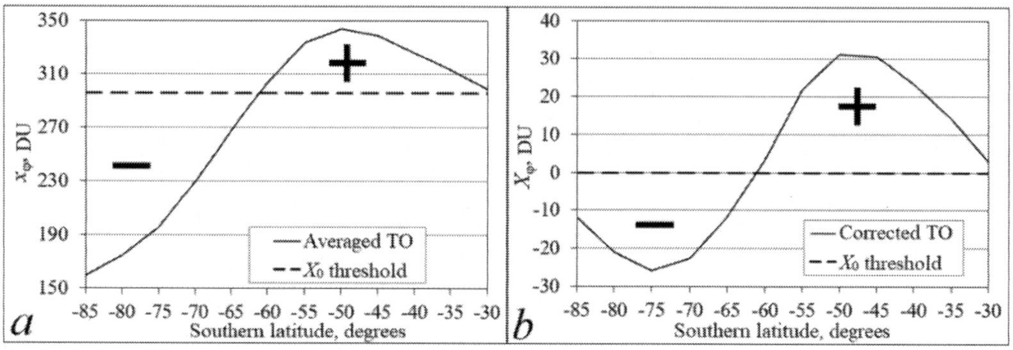

Fig. 3. Deviation of ozone content from the X_0 threshold in the polar and middle latitudes of the Southern Hemisphere.

In Figure 3,*a* it can be seen that the areas «–» and «+» differ in size. Here the spherical shape of the Earth is not taken into account. The zonal TO should be multiplied by the cosine of latitude φ for sphericity correction: $X_\varphi = x_\varphi \cos \varphi$, X_φ is corrected value of TO.

Denote deviation of the corrected value of X_φ from the threshold as $\Delta X_\varphi = X_\varphi - X_0$.

Dependence of ΔX_φ values on the latitude is given in Figure 3,*b*.

Sum up deviations from the threshold X_0 ΔX_φ separately with for the area with «–» sign and for the area with знаком «+» sign. In the first case the sum of ΔX_φ values specifies ozone deficit *D* in the polar area. In the second – the sum of ΔX_φ values for the area with «+» sign it is excess of ozone *E* in the middle latitudes.

Results of zonal TO studies for 1985-2015 are presented in the Table. The first column is the years, the second – threshold X_0, deficit *D* is the third column, excess *E* – in the fourth. About the difference *Q* of deficit and excess one can judge from the last column which shows this difference in per cent. Value *Q* is calculated by formula:

$$Q = \frac{D-E}{(D+E)/2}.$$

Table

Year	X_0, Du.	*D*, Du.	*E*, Du.	*Q*, %
1987	305.7	151.8	153.6	-1.2
1989	312.9	149.1	152.4	-2.2
1990	304.4	138.1	135.4	2.0
1991	311.2	130.5	203.5	-43.7
1999	299.2	161.5	151.6	6.3
2005	302.3	128.5	130.4	-1.4
2010	295.7	121.7	115.9	4.9
2013	295.5	89.7	98.1	-8.9
2014	302.8	123.6	129.8	-4.9
2015	296.5	198.3	94.5	70.9

The estimates given in the Table show that the deficit of ozone *D* in the polar areas and its excess *E* in the middle latitudes practically coincide, and value *Q* is of the same order as the measurement error.

The exception are years 1991 and 2015, when *Q* was -43.7% and 70.9%, respectively. This seems to be associated with the action of Pinatubo (1991) and Calbuco (2015) volcanoes. During volcanic activity eruption products (e.g. sulfur dioxide) are emitted and can destroy the ozone layer [8]. Eruption of Pinatubo volcano (15.07°N; 120.21°E) occurred in the Philippines in 12-17 June1991 was with Volcanic Explosivity Index VEI = 6. Eruption of Calbuco volcano (41.33°S; 72.61°W) with VEI = 4 occurred in Chile 22-30 April 2015. These events might partially deplete the ozone in the Southern Hemisphere and result in the mismatch of the ozone deficit in the polar regions and its excess in the middle latitudes.

Regretfully we did not succeed to analyze the matching of ozone deficit and excess for every year. The main case was partial or complete lack of zonal data on TO. For 1993 – 1995 these data are completely absent. For 1985, 1986, 1992, 1996, 1997, 1998, 2002, 2003, 2007, 2008, 2011 omissions in the zonal data up to 33 days. This denied correct choice of ΔT time interval averaging and defining the threshold X_0.

5 Conclusion

Methods of evaluating the ozone deficit in the polar areas and its excess in the middle latitudes of the Southern Hemisphere during existence of the Antarctic ozone hole have been developed. The essence of the method is to compare the amount of ozone in the polar and middle areas with a certain threshold value. Mean annual total ozone content in the latitude range from 32.5° to 52.5°S for every year was experimentally found to be the threshold value. Estimates have been made to show that during the existence of the Antarctic ozone hole the ozone deficit in the polar areas and its excess in the middle latitudes of the Southern hemisphere practically coincide. Estimates of the difference between the deficit and the excess in most cases are of the same order as the measurement error.

Our studies confirm the results of R.S. Stolarski and M.R. Schoeberl and indicate that the Antarctic ozone hole is a natural geophysical formation. We believe that the theory of the origin of the hole from chemical ozone depletion is not correct.

References

1. J.C. Farman, B.G. Gardiner, and J.D. Shanklin, Nature, **315**, 207 (1985)
2. NASA Ozone Watch, Images, data, and information for atmospheric ozone, http://www.ozonewatch.gsfc.nasa.gov
3. R.S. Stolarski, M.R. Schoeberl, Geophys. Res. Lett., **13,** 1210 (1986)
4. S. Solomon et al., Nature, **321,** 755 (1986)
5. S. Solomon et al., Science, **353**, 269 (2016)
6. NASA, Ozone, ftp://jwocky.gsfc.nasa.gov/pub/
7. V.B. Kashkin, T.V. Rubleva, and R.G. Khlebopros, *Stratospheric ozone: space orbit view* (Krasnoyarsk, SFU, 2015)
8. V.V. Zuyev, N.Ye. Zuyeva, Optics of the atmosphere and ocean, **24,** 30 (2011)

Framework of cloud web services for processing remote sensing data

Alexey Buchnev[1], *Pavel* Kim[1], *Valery* Pyatkin[1,*], *Fedor* Pyatkin[2], and *Evgeny* Rusin[1]

[1]The Institute of Computational Mathematics and Mathematical Geophysics SB RAS, Academician Lavrentyev Ave., 6, Novosibirsk, 630090, Russia
[2]Siberian Centre Scientific Research Centre of Space Hydrometeorology 'Planeta', Sovetskaya St., 30, Novosibirsk, 630099, Russia

Abstract. We consider a distributed network of cloud web services for processing satellite data, which provides data processing facilities for Earth remote sensing within SaaS model. In fact, this is a set of web services that implement the functional modules of the PlanetaMonitoring remote sensing data processing system.

Introduction

The aim of this work is to develop and implement a model (prototype) of a framework for distributed network of cloud web services for processing satellite data, which provides data processing facilities for Earth remote sensing within the SaaS model. In fact, this is a set of web services that implement the functional modules of the PlanetaMonitoring system for thematic processing of remote sensing data [1].

1. PlanetaMonitoring software

PlanetaMonitoring software was developed as a joint effort of the Institute of Computational Mathematics and Mathematical Geophysics (ICM&MG) SB RAS and the State Research Centre 'Planeta'. It provides a functionally complete set of operations for processing remote sensing data and implements a number of software technologies for processing multispectral satellite information in the optical, infrared, and microwave ranges. Here are some of these software technologies: satellite image filtering; radiometric and geometric correction; geo-referencing; transformation into cartographic projections and the construction of mosaics from individual images; detection of lineaments and ring structures; recognition and classification of environmental objects (cluster analysis and supervised classification); determination of objects` spatial movements using multi-temporal satellite images. To use the above-listed parts of PlanetaMonitoring in the cloud environment, we implemented each of them as autonomous (i.e. non-interactive) Windows. This article proposes an approach to the organization of a distributed cloud environment capable of ensuring reliable and efficient execution of these Windows applications when processing remote sensing data.

* Corresponding author: pvp@ooi.sscc.ru

2. Cloud computing

At present, the use of Internet technology for the operational integration of information and computing resources in solving the problems of processing remote sensing data is becoming relevant. The paradigm of cloud computing gives such an opportunity [2]. The main idea of cloud computing is distributed data processing and storage technologies, where all necessary resources are provided to the user as an Internet service. There are the following kinds of cloud computing technologies, Infrastructure as a Service (IaaS), Platform as a Service (PaaS), Software as a Service (SaaS), and many other. In ICM&MG SB RAS, we develop the prototypes for web services that use software technologies (autonomous Windows applications) of the PlanetaMonitoring software package; which solves a new problem of providing remote sensing data processing facilities within the cloud SaaS model [3]. Service prototypes are implemented on the Windows platform and consist of the following two components:

- *computational component*, created on the basis of the a previously developed corresponding Windows application (autonomous mode); and
- *web interface*, based on the free and open-source Apache web server.

The main ('master') cloud web server for satellite data processing is implemented on the basis of ICM&MG SB RAS. A number of circumstances impose such a solution. First of all, a so-called cloud environment is formed on the main server, providing remote sensing data processing services as part of the SaaS cloud model [4]. It is important that there is the possibility to organize high-performance scalable computing at the Siberian Supercomputer Center (SSCC) SB RAS, which is a part of ICM&MG SB RAS [5].

3. High performance scalable computing

The developed framework assumes the use of high-performance SSCC resources for time-consuming calculations. The high-performance computing subsystem is implemented on the basis of the SSCCIP (Siberian Scientific Computing Center – Image Processing) software developed by the authors for integrating remote high-performance computers into the processing and analysis of satellite data [6]. SSCCIP consists of four components:

- *client component* - the operator's workplace in the Windows environment (it provides a GUI for configuring a request for data processing, tracking its execution results, and visualizing them);
- *server component* - Unix-application running on a supercomputer (it provides the processing of the client component request and its dispatching);
- *communication component* - network exchange between client and server components (which is implemented on the basis of the secure SSH protocol);
- *computational component* – on the basis of the high-performance image processing library ParImProLib [7] developed by the authors. The library provides calculations on the NKS-30T + GPU SSCC hybrid cluster: inter-node exchange is provided by the MPI interface; and calculations on the GPU (currently in development), by Nvidia CUDA technology. In the future, it is planned to expand the library with the function of processing on the Intel Xeon Phi CPU to enable it to perform calculations on the new SSCC cluster NKS-1P.

Note that the SSCCIP system and the ParImProLib library are also built in the form of a framework: their source code is organized so that operations common to a certain type of processing algorithms are separated from the operations specific to particular algorithms of this type. In [6], it was shown that this approach allows maintaining high performance of the program code while significantly simplifying the process of adding new processing algorithms to the system. To use SSCCIP as a subsystem of a cloud web service, its client component will be enhanced with the ability to perform autonomous processing with an interface identical to the interface provided by PlanetaMonitoring. Thus, the framework being developed will isolate the web server from the special characteristics of a particular computer.

4. Roshydromet Core Ground Segment (CGS) Ground system of receiving, processing, archiving and distribution of satellite data at State Research Center 'Planeta'

The cloud environment implemented on the main server (either totally or partially) is transferred to the CGS servers using the "mirroring" technology [8]. The Ground system of receiving, processing, archiving and distribution of satellite data includes three regional centers that are the part of the State Research Center 'Planeta': European (Obninsk – Moscow – Dolgoprudny), Siberian (Novosibirsk) and Far Eastern (Khabarovsk) ones. In addition, the CGS includes a network of fixed and mobile autonomous satellite data reception points (about 70) in Russia, Antarctica, and on marine vehicles under the scientific guidance of the State Research Centre 'Planeta'. The first backup mirror will be implemented on the server of the Siberian Center (SC) of the Scientific Research Centre 'Planeta'. It is important that this Scientific Research Centre supports the web server, which presents the operational satellite data arriving in real-time mode (ftp1.rcpod.ru).

5. Computational components of a distributed cloud web services system

Virtually every computational component of a web service is an autonomous version of the corresponding software technology of the PlanetaMonitoring system. This article presents a cloud ptototype consisting of two web services created on the basis of previously developed Windows applications: LINECOIL (lineaments and ring structures' detection on satellite images) and KMEAN (clustering by the K-means method). To access the corresponding web service, one need to transfer a text file with processing parameters to the cloud environment.

Note that the technology of detecting lineaments and ring structures has already been implemented in the form of a cloud web service and is available to the Internet users [3]. At present, the testing of a web-based cloud clustering of remote sensing data is being completed. The main clustering algorithm in our system is the known K-means algorithm [9]. The algorithm is based on an iterative procedure for assigning feature vectors to clusters by the criterion of the minimum vector distance to the center of the cluster. It is considered optimal to divide the input vectors into clusters, at which the intraclass spread cannot be reduced when transferring any vector from one cluster to another. The K-means algorithm provides the minimum (in the general case of the local) of the following error function:

$$E = \sum_{i=1}^{K} \sum_{j=1}^{N} \gamma_{ij} \rho(x_j, m_i), \gamma_{ij} = \begin{cases} 1, x_j \in C_i \\ 0, x_j \notin C_i \end{cases}$$

Here N is the number of feature vectors, x_j is the feature vector number j, K is the number of clusters, C_i is the cluster with number i, m_i is the center of the cluster number i, $\rho(x, y)$ is the selected metric (distance) between the vectors x and y.

The results of clustering can be controlled using the following parameters [10]: the number of clusters to be allocated; the number of iterations of the algorithm; metric type (distance) between vectors (Euclidean, Chebyshev, Mahalanobis, Manhattan), the selected metric defines the shape of the resulting clusters; the method of selecting the initial cluster centers (one out of three, two of them are determined on the basis of the data set statistical characteristics and one is based on a random sampling); and computing accuracy. In addition, it is possible to save statistical data of the obtained cluster map (cluster volumes, vectors of mean values in clusters, standard deviations, etc.) in a text file.

6. Web interface of a distributed cloud web services system

The development of "private" cloud services has also such an argument as the possibility of ensuring the safekeeping of intellectual property for the creation and use of proprietary software algorithms in the know-how mode. Thus, it turns out to be justified to invest some effort into the cloud technologies development, which is incommensurable in scale to the capabilities of software corporations. When designing their software content, the numerical expansion of the web services range (developed within the framework of a single cloud technology) should take into account qualitative features in the structure of user requests, in particular - their async origin. Large time expenditures for servicing a particular online service require tracking its current state, and the decision to interact with the user on this task successfully combines the balance between the computational loads on the process itself and its auxiliary software. Another vital problem of cloud development, the solution of which increases its effective potential, is the processing of multiple concurrent same-type requests. It can be most possible without serious consequences when "mirroring" service processes with the help of an available network, i.e. duplicating infrastructure using file system directory hierarchy repositories that fit into the hypertext space of possible web services browsers. The "mirroring" operation is provided by the resource allocation manager for specific service requests. The architecture of embedding the object-oriented cloud of web services into the distributed CGS network determines the structural complexity of the cloud resource allocation manager. "Replaceable" computational components rely on a unified network interface and provide for the possibility to implement computational components on various architectural platforms of heterogeneous networks, combining both multiprocessor SSCC architectures and CGS distributed networks. Considering the heterogeneity and vastness of the market of user browsers used on the Internet, the web interface in the prototype version is implemented using the basic means of the hypertext language HTML. The service part is made on Apache 2.2, while taking into account the natural heterogeneity of the underlying infrastructure, other platforms are being also developed in the IIS Windows architecture. Informational data streams are provided via FTP protocol. Thus, the user has the ability to search for "hidden" in the "cloud" of objects, both in archival data and in individual samples of images. The cloud storage paradigm is tracked here.

Fig. 1-4 show an example of a session with a K-means cluster web service prototype: Fig. 1 is a screenshot of the configuration web interface form setting up the detection of 15 clusters; Fig. 2 is the the source image of the flood in the Tyumen region (satellite Sentinel-2, resolution 10 m, 05/11/2017); Fig. 3 is the result of clustering using the Euclidean metric; Fig. 4, the Mahalanobis metric.

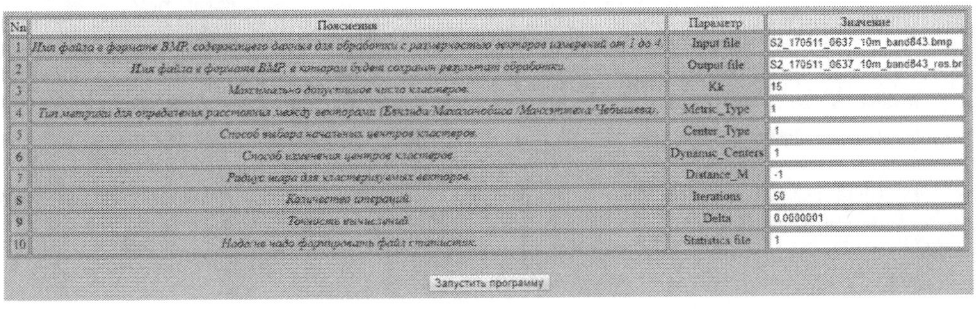

Fig. 1.

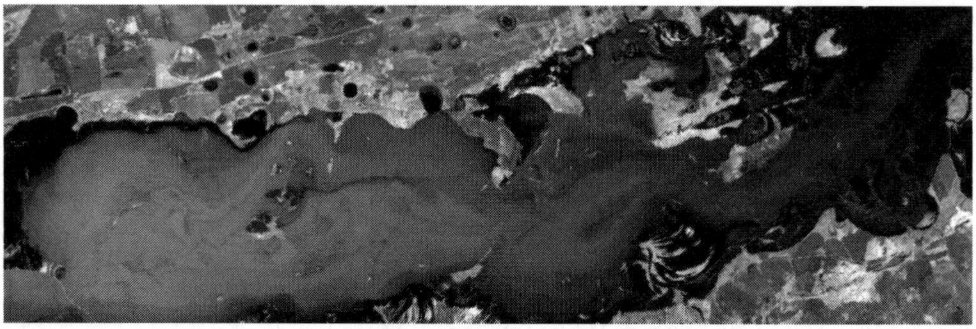

Fig. 2.

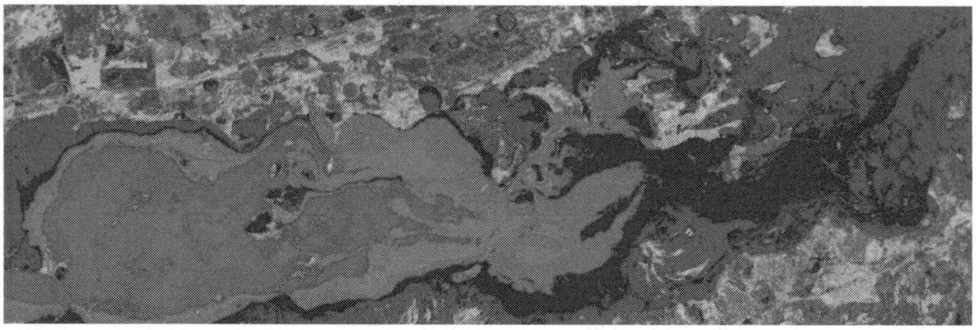

Fig. 3.

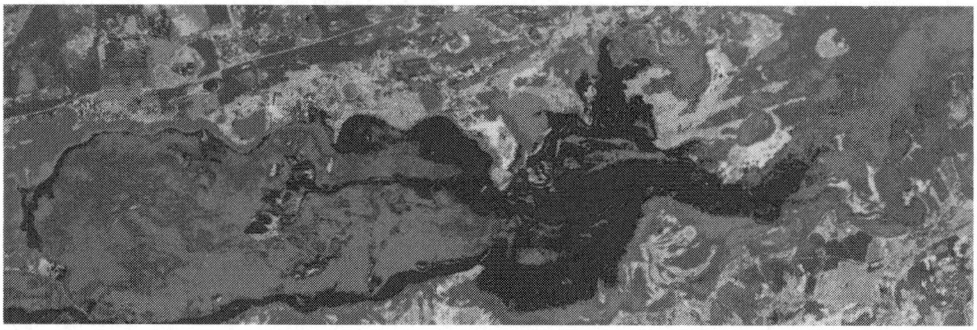

Fig. 4.

Conclusion

Successful experience in implementing a cloud web service prototype for detecting lineaments and ring structures on space images was used to create the cloud service for remote sensing data clustering, which represents the known K-means algorithm. In the future, it is planned to develop the prototypes of cloud services for other programming modules of the PlanetaMonitoring system.

The work was conducted within the framework of the budget project 0315-2016-0003 for ICM&MG SB RAS with the support of Russian Foundation for Basic Research (grant 16-07-00066). Siberian Supercomputer Center SB RAS is gratefully acknowledged for providing supercomputer facilities.

References

1. V.V. Asmus, A.A. Buchnev, V.A. Krovotyntsev, V.P. Pyatkin, G.I. Salov, Optoelectronics, Instrumentation and Data Processing **54(3)** 1–9 (2018)
2. P.A. Kim, P.A. Kalantaev, V.P. Pyatkin, Pattern Recognit. Image Anal. **25(2)**, 220–222 (2015)
3. A.A. Buchnev, P.A. Kim, V.P. Pyatkin, G.I. Salov, Journal of Siberian Federal University. Engineering and Technology **10(6)**, 741–746 (2017)
4. A.A. Buchnev, V.P. Pyatkin, F.V. Pyatkin, Interexpo Geo-Siberia, 8–13 (2017)
5. SSCC ICMMG SB RAS, http://www.sscc.icmmg.nsc.ru
6. E.V. Rusin, LNCS **6873**, 467–472 (2011)
7. E.V. Rusin, LNCS **5698**, 344–349 (2009)
8. Mirror website, https://en.wikipedia.org/wiki/Mirror_website
9. A.K. Jain, Pattern Recognit. Lett. **31**, 651–666 (2010)
10. V.V. Asmus, A.A. Buchnev, V.P. Pyatkin, Optoelectronics, Instrumentation and Data Processing **46(2)**, 149–155 (2010)

Technology of 3d relief modelling based on Delaunay triangulation algorithms

Gachenko Andrey[1,1] *Khmelnov* Alexey[1]

[1]Institute for System Dynamics and Control Theory of Siberian Branch of Russian Academy of Sciences, Lermontova St., 134, Irkutsk, Russia

Abstract. This paper represents the algorithm of superimposing underwater relief and land relief basing on heterogeneous initial data with the use of Delaunay triangulation. As a result of superimposition a qualitative 3D relief model has been created. This model can be used in generating forecasts of reservoir shoreline alterations related to hydroelectric power plant flushing. The described technology has been approbated in various practical tasks and has shown its efficiency.

In the course of flooded area calculation in the lower pond of the Irkutsk HPP there appeared the necessity of creating the relief model which includes the river bottom contour. The land relief being represented with different sophistication on topographic vector electronic maps in various scales, no information on the underwater relief of the Angara River could be found. The existing soundings information for certain navigators is only presented fragmentally, and, according to specialists, is not accurate. The only reliable source was the atlas [1], which was distributed in paper form. It brought up the necessity of digitizing the maps presented in the atlas and superimposing the maps on the available topographic base.

The digitizing of the paper atlas was made with the use of Easy Trace 7.99 shareware vector map generator. Depth contour lines, depth marks, shoreline have been digitalized.

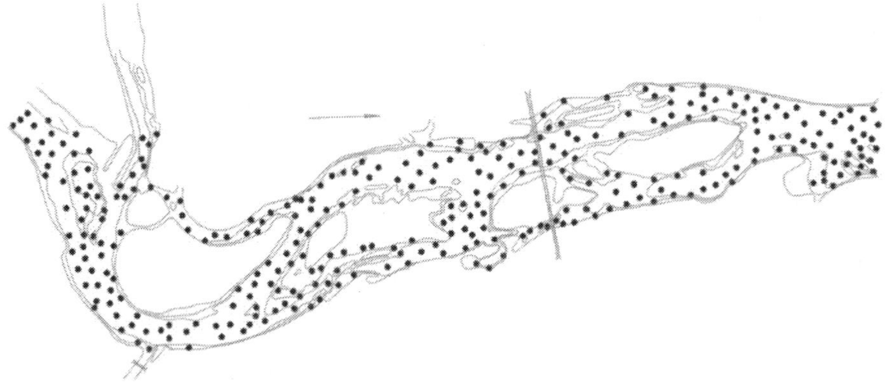

[1] Corresponding author: gachenko@icc.ru

Fig. 1. Result of digitizing

The atlas sheets are oriented in arbitrary directions in order to place the reaches of the river on the pages in a more compact way. For the future use of the atlas sheets a rotation and a shift into the relief model map coordinate system have been performed. To choose rotation parameters, the distinguished points (capes, bays, island top ends) of the two maps have been juxtaposed. Supporting software has been designed to calculate the transformation parameters from the given list of positioning data pairs and to transform the map. The least square method is used to search for the transformation parameters. The method requires only two pairs of corresponding points, but the use of a bigger number of points allows obtaining more accurate results as well as evaluating the accuracy by standard deviation. In this respect, the standard deviation is 25-70 m. The deviation value is accounted for by the significant contour divergence, as the divergence may touch upon the distinguished points used for the juxtaposition of the maps.

Fig. 2. Juxtaposition of the maps

An attempt to superimpose the contours of the vectorized shoreline on the shoreline of the topographic base has shown that the atlas was created schematically, without a reference to any map (Fig. 2). That is, the atlas shows distinguished shoreline curves, but it is impossible to superimpose this shoreline on a more accurate contour by means of rotation and shift. The experience of the use of the rotation and shift transformation calculated from the juxtaposed distinguished points has demonstrated that the obtained result cannot be significantly improved by means of making transformation parameters more precise.

In order to make it possible to use insufficiently accurate data to create a relief model, software for electronic map morphing has been developed. To perform morphing it is necessary to find continuous plane transformation which can superimpose inaccurate shoreline contours on more accurate ones. The layers of the shorelines are used to superimpose the maps because they are present on both maps. Thereafter, the same transformation is applied to other layers of the underwater relief (isobaths and depth marks). The obtained data accord better with the land relief data, at least the transformed isobaths and depth marks do not lap over the shoreline. The transformed shoreline contours are not used as they are, as more accurate shorelines are already available, that is why the target line deviation is excusable, for example, as a result of adding superimposition interdiction. The crucial thing is that contour deviation areas should not contain the data of the used layers of the inaccurate map. To perform map morphing two separate software modules are designed: the first one is for semi-automatic parameter setting for contour superimposition and the second one is for performing transformation.

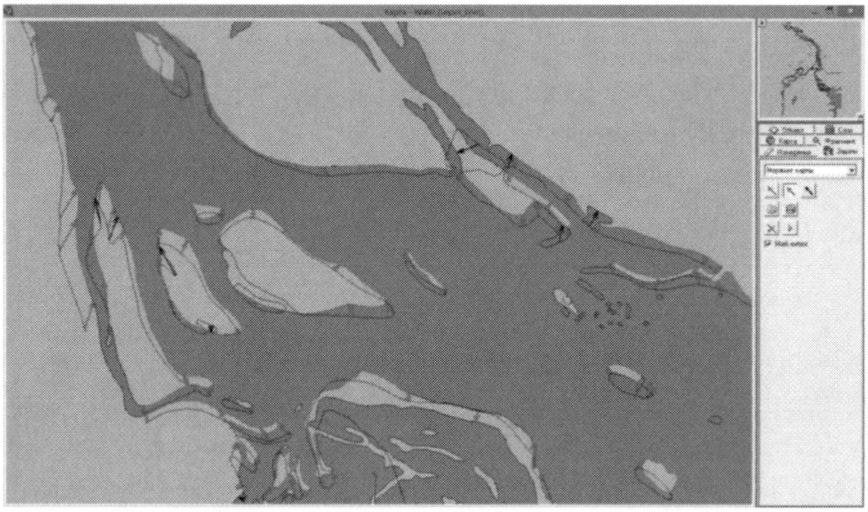

Fig. 3. Transformation adjustment for contour superimposition. The accurate map is shown with painting, the transposed map is in contour representation

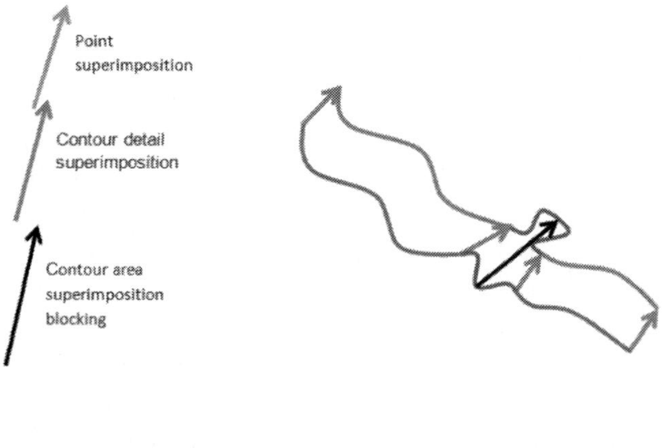

Fig. 4. Map superimposition operations

The map superimposition is performed as a special task with the help of IrkGV map viewer (Fig. 3). The task constitutes in setting arrows which superimpose the points of the original (inaccurate) map and the target-oriented (accurate) map. Three types of arrows are used to superimpose significant points, to superimpose contour points, and to interdict the superimposition of contour details (Fig. 4). When superimposing contours, the initial point and the finishing points of the arrows drawn by the user are projected into the closest contour points of the corresponding maps. Next, we draw display of every contour detail of the original map object into the specified contour detail of the target-oriented map object. It is done for each pair of arrows which are closely spaced on both contours. Thus, if we, for instance, add an arrow between the points of one of the contours which superimposes this point on the third contour, there will be no superimposition of the contour details.

Juxtaposition of two closed contours is to be processed separately. Without taking it into account, we would have to set no less than three pairs of correspondent points in order to bind the contour details. When choosing details to juxtapose by a smaller number of correspondent points, contour tracing direction is to be taken into account: the contour details which have identical contour tracing are to be juxtaposed. It allows setting correspondence between contours using only one pair of points.

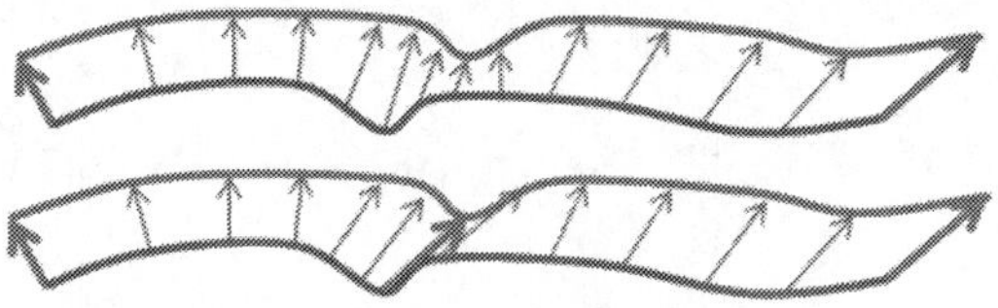

Fig.5. Adjustment of distinguished points superimposition by means of adding their obvious juxtaposition.

The display of the points that belong to the juxtaposed contour details is performed with the help of the linear transformation of parametric curve coordinates. If the result is not satisfactory, one has to clearly juxtapose the distinguished points of the contours to ensure juxtaposition of the contours (Fig. 5). After imputing data on the superimposition of the maps a transposition file is formed which includes not only obviously set shifts, but also the shift that have been calculated for temporary points of the contours. Superimposition interdiction is required in cases when the juxtaposed fragments are represented with different sophistication. For example, the accurate contour depicts a bay which is not present on the less accurate one. In this case, one had better avoid juxtaposing the fragment correspondent to the bay and the straight part of the trasnformed contour (Fig. 4).

Morphing was executed with the help of triangulation building software [2]. In doing so, limited triangulation is built by the initial points from the transformation file. Ties are added for the nearly points of one contour. Shift vector to the target point is memorized for each triangulation point. Then, layer processing in Shift format is performed: each point is transposed onto the vector which was obtained by linear interpolation from triangulation points.

To juxtapose the data on the underwater relief and the land relief, map layer triangulation building form has been improved. The role of each layer in triangulation building is now specified. The layer may contain the following information: land relief, standing water level, shoreline contour, underwater relief. Underwater relief operations require standing water level information. These data are used to build subsidiary triangulation which provides hypsometric data. The shoreline contour, in regard to which the depth is measured, is linked to these data.

Isoline contouring by triangulation is used to calculate the flooded area. The height of isoline points are determined by the difference of height of the original relief and the calculated water level after flooding. This approach allows obtaining more accurate results and covering a bigger area in comparison with commonly used isoline contouring by terrain model for the height equal to the water level at a certain point increased by the height of water lift at this point.

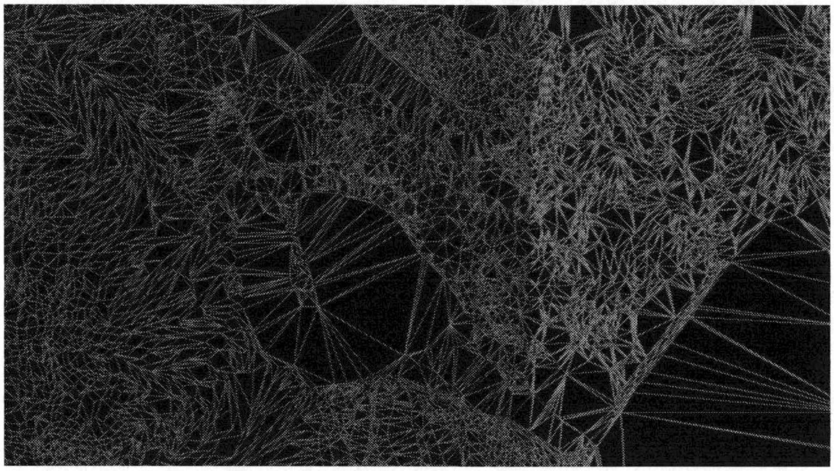

Fig 6. A fragment of built triangulation containing information on land and underwater relief

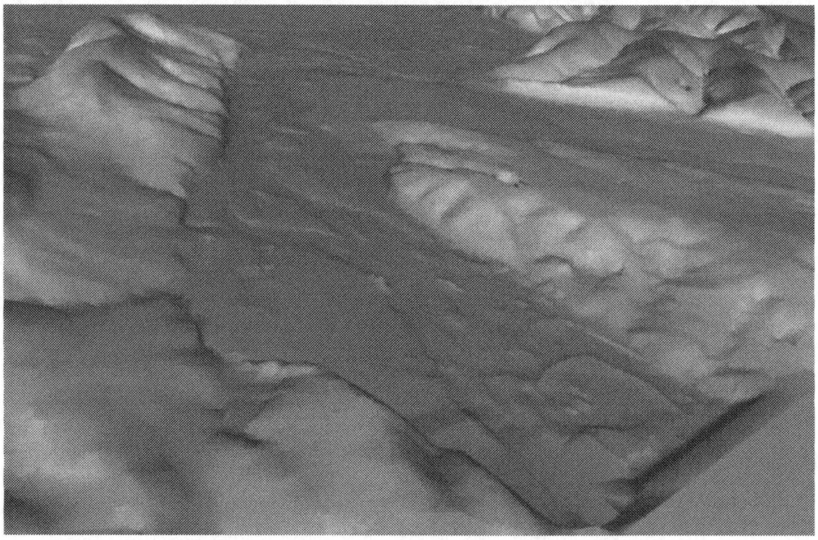

Fig. 7. Superimposed relief (the picture is blown up for clarity)

The created relief model is saved in the triangulation file (.trg files) and can be used for future analysis with the application of TrigLib.dll dynamic library designed to deal with these data. This library allows deriving relief height from the triangulation at an arbitrary point and tracing the line, i.e. to obtain contour interval along a fragment or a curve.

Apart from principal triangulation, which contains relief data, subsidiary triangulation is formed, being built by standing water level, which contains data on water level surface used to create the general relief model.

Triangulation building program implements the algorithms for river valley cross section building by the coordinates specified in the file, relief contour building by specified triangulation, as well as subtracting heights specified by one triangulation from the height points of another triangulation. Thus, the triangulation based on the calculated water level is built to plot flooded areas.

Using the TrigLib.dll dynamic library, the software for altimetric data generation by arbitrarily given contour of the Angara River has been elaborated.

Conclusion

As a result of extreme flooding evaluation under various cases of extreme water content in the basin of the Angara River and Lake Baikal the hybrid information geographic system has been created, which allows simulating various flooding and shallowing scenarios and determining flooded zones with account of underwater and land relief. The described technology has been approbated in various practical tasks and has shown its efficiency.

References

1. *The map of the Angara river from Irkutsk HPP to 142 km* (the Ministry of Transport of the Russian Federation, River Transport Department, Glavvodput', "Water Routes of the East-Siberian basin" state owned enterprise, Irkutsk, 1994)
2. A.V. Skvortsov *Delaunay triangulation and its application* (Tomsk: Tomsk University Publishing House, 2002)

Complex for reception and real time processing of remote sensing data

Oleg Yakubailik[1,*], *Victor* Romas'ko[1], and *Evgeny* Pavlichenko[1]

[1] Federal Research Center "Krasnoyarsk Science Center of the Siberian Branch of the Russian Academy of Sciences", 50/45 Akademgorodok, Krasnoyarsk 660036, Russia

Abstract. The basic problems and trends in the development of modern systems for the reception, storage and real-time processing of satellite data are considered. Abrupt increase in the capability of satellite systems, significant increase in the amount of satellite information and its availability, the development of data processing and presentation technologies, and the use of web technologies are discussed. Data sources of modern remote sensing systems of the Earth and the features of their practical use are considered. It is concluded that the most effective way to obtain real-time information from meteorological satellites are satellite stations that receive data in the X-band at a frequency of 8 GHz. The performance characteristics and capabilities of the equipment of the new satellite receiving complex at Krasnoyarsk Science Center are given. Use of up-to-date computer equipment (high-performance servers and storage systems, local area network with a bandwidth of 10 Gbit/s) and logical separation into the stages of data conversion (data reception, primary and thematic processing) provide the construction of a modern scalable data-processing system for remote sensing data. The paper presents the results of the work on creation of specialized software for information and analytical systems for real-time satellite monitoring.

1 Introduction

The geographical features of the Krasnoyarsk Region (extensive territory with a diverse relief and climate, high proportion of poorly developed and inaccessible zones, large reserves of natural resources and other factors) objectively lead to the need for the use of Land remote sensing systems (RSS) and various thematic information products based on them. Integration of the RSS data with real processes of providing vital functions to public authorities and the population assumed major importance of a strategic factor for further accelerating the social and economic development of the Krasnoyarsk Region.

One of the stages in providing the effective use of space activities in the Krasnoyarsk Region was purchase and put into operation of a new satellite receiving complex FRC KSC SB RAS on the basis of UniScan-36 station by ScanEx Co. in 2017. This complex provides the receiving RSS data from all the relevant low-orbiting meteorological satellites of medium spatial resolution (TERRA / AQUA, Suomi NPP / NOAA-20, FengYun-3).

A large number of information systems based on earth remote sensing data have been developed in the world [1-5]. We can assume that the general arrangement of the RSS data processing has already been sufficiently established; basic units include receiving and archiving, as well as primary and thematic processing [6]. At the first stages of monitoring systems development all the stages of processing remote sensing data had to be implemented in the construction of a specific system and a special infrastructure had to be completely created. But today the use of specialized centers for data storage, processing and analysis is more efficient and appropriate from different points of view [7]. It should be noted that significant changes are currently taking place in the area of applying RSS satellite data in solving various scientific and applied problems as follows [8-10]:

- The capabilities of RSS satellite systems have abruptly increased. It became possible to organize monitoring of rapidly occurring processes due to the increase in the number of RSS satellites. Over the past 4 years, the number of satellites has increased by an order of magnitude, the availability of high-resolution data has increased (up to 3 m/pixel).
- The number of RSS satellite systems with "measuring" properties has increased. Such systems provide not only qualitative but well-calibrated quantitative information about various objects, processes and phenomena.
- The level of information availability has increased. The data of many satellite systems began to be freely expanded both through the Internet and directly from satellites in an unencrypted form.
- A significant increase in the amount of satellite data and the development of the requirements and cost of the systems for their reception and primary processing led to the intensification of transition from the use of personal RSS processing systems to service-oriented solutions based on regional specialized remote sensing centers.

Along with the extensive development of RSS satellite systems in the last decade, there has been a fairly abrupt increase in the amount of satellite data received from individual spacecraft, as well as the increase in the nomenclature and total amount of information products generated on the basis of this data processing [11-13]. Thus, the growth rate of information amount exceeds the growth rate of the number of remote sensing systems. For example, the volume of RSS data archives in NASA and NOAA, accumulated over the year, increased by more than an order of magnitude: from almost 1 Petabyte in 2000 to more than 10 Petabytes in 2011, although the number of RSS satellites increased during this time not more than in 6 times. Entering new RSS data into the archiving system in 2018 according to NOAA plans will be 20 Petabytes per year, and their total volume will be more than 100 Petabytes. In this case, the amount of data per satellite will increase by 100 times. Similar plans to build up RSS grouping and a rapid increase in the amount of RSS data were accepted for implementation in Russia (the Federal Space Program of Russia for 2016-2025).

Thus, we should mention here the abrupt increase in the capabilities of satellite systems, a significant increase in the amount of satellite information and its availability, the development of data processing and presentation technologies, and the use of web technologies [14-17].

2 Data sources of modern remote sensing systems

The equipment of modern remote sensing systems is traditionally built around the main sources of remote sensing data. RSS data can be obtained 1) on request from the official RSS data providers (usually high spatial resolution data), 2) via the Internet from free sources and, finally, 3) direct reception of data from satellites operating in broadcast mode. Free information of low and medium resolution, and recently also high resolution, mostly is available for download from official resources, such as the USGS Earth Explorer service of the US Geological Survey [18]. This way for obtaining data in our country is gradually assuming increasing importance as Internet access becomes faster and cheap. However, not all the satellite data is available through such services, especially for new and developing systems. Another problem is delays in data processing. They are not always processed quickly enough. This factor can become important for tasks such as real-time monitoring of floods or fires. An unpleasant theoretically possible prospect for such services is the introduction of a fee for use, which can completely change the principles of financial support for the processing of remote sensing data.

The analysis showed that the most efficient and reliable option for obtaining medium and low resolution data is one's own (or partner) station for receiving space information. Among the existing standards of data transmission in the ranges of 137 MHz, 1.7 and 8 GHz, the latter one allows to transfer the maximum amount of information per session. Such a station is capable of receiving up to 320 Mbit/s depending on the diameter of the antenna, the number of reception frequencies (receivers) and the type of phase modulation [19].Taking into account the orbital altitude of the polar-orbiting satellites of about 800 km and the capture band of 2-3 thousand km, the viewing area of such a station represents approximately an ellipse measuring 5 by 8 thousand km long in longitude. For the station located near the geographical center of Russia, in Krasnoyarsk, about 80% of the territory of Russia is viewed twice a day from each receiving device. Such a station of X-band frequencies (8 GHz) forms the basis of the satellite receiving complex of FSC KSC SB RAS.

3 Organization of satellite data processing

The hardware-software arrangement of the RSS data processing in case of having its own satellite data reception station consists of 3 logical groups of servers – receiving servers, preliminary processing servers and thematic processing servers. It includes also workstations for interactive thematic processing (Fig. 1). Data storage systems for remote sensing data, characterized by large daily and session volumes, can be integrated into the servers of these logical groups or be located separately, perhaps as part of a larger infrastructure. The division of servers and processes into these three groups logically follows from the characteristics of each group of receiving and processing.

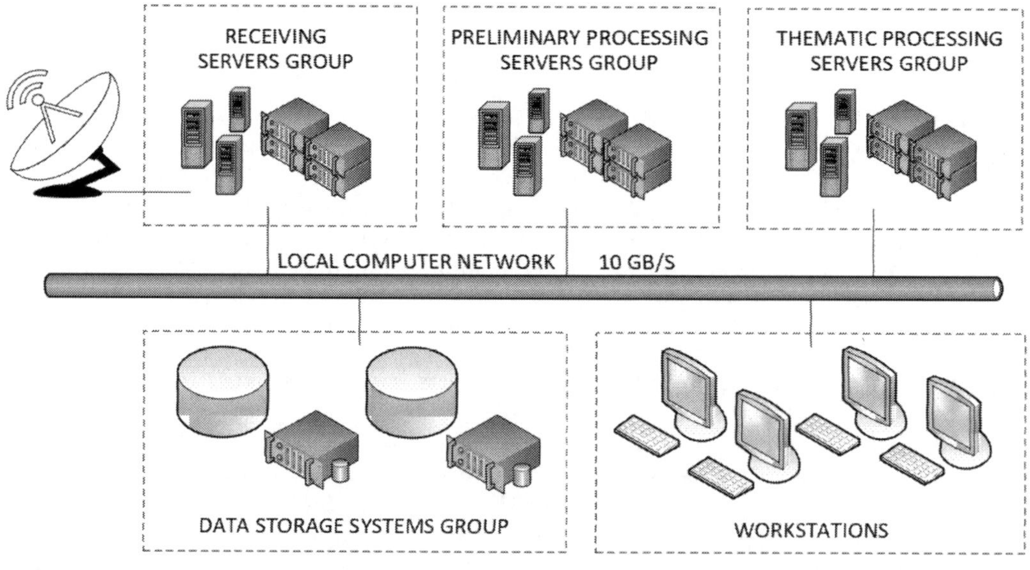

Fig. 1. Hardware-software arrangement of the receiving satellite complex at FCC of Krasnoyarsk Science Center of the Siberian Branch of the Russian Academy of Sciences.

The group of data receiving servers is allocated for the reason of working in real time mode at the moments of data reception by the radio channel and therefore is not loaded with any additional processing loads. The task for the servers of this group is to accept and to record on the disk memory the entire data stream from all sessions without loss.

The preliminary preprocessing servers group includes the processes of gridding and radiometric calibration of the received data. It is logically isolated because preliminary processes are required for all the space RSS and are performed firstly. Moreover, they are implemented based on the software specific for each RSS that is provided by the developers of this system.

The third group of thematic processing servers, in fact, fulfills the target data processing with obtaining the final useful information for consumers. This group includes thematic software of FSC KSC SB RAS own design. Processing here is more homogeneous and consistent, and also most often change due to the development of new versions or applications. This group includes a web interface built on the basis of geo-portal technologies, as the basic access point for consumers to remote sensing information [20-22].

The satellite reception complex includes the following equipment:

- The receiving station UniScan-36 manufactured by SCANEX Company [19] includes a parabolic antenna with a diameter of 3.1 m, a three-axis supporting-rotating unit, two channels of data reception and recording based on the rack-mountable PC, and stationary GPS receiver. The antenna is installed on the roof of the building, on a specially constructed metal structure site. The GPS receiver is mounted on the guard. A rack with receiving equipment is located in the room on the top floor, directly below the antenna, since the maximum length of the high-frequency cable of the antenna is 50 m.
- Data processing servers (2 pcs) by HP Company, one of which is responsible for the operational real-time primary data processing (Level 1), the second is designed for in-depth thematic processing and solution of applied problems.
- Clustered storage systems (2 pcs) based on hybrid disk arrays Network Appliance FAS2554 [23] (each of them includes 20 HDDs per 4 TB + 4 solid state drive SDD for data caching. The available capacity of the fault-tolerant storage is 36 TB).
- High-performance workstations (4 pcs).
- Local area network of 10G standard with data transmission at rates up to 10 Gbit/s through the copper twisted pair, built on the basis of the 16-port switch Extreme X620-16T [24],which provides full compatibility with less rapid standards. Cable lines in 10G standard are limited by a length of 50 m and require a cable and "twisted pair" connectors of 6 category or higher. Practically, the line with length of 48 m showed an exchange rate according to the protocol CIFS2 between the workstation and the server during 10 GB file copying in the beginning – 1100 Mb/s, and at the end – 190 Mb/s, which corresponds to the real recording rate on the disk array. On the line with a length of 53 m the adapters already automatically switched to the 1G standard. Thus, a modern compact local computer network is many times surpassed any disk array according to its throughput, even on the basis of solid-state devices, and is no longer a "narrow throat" of the system.

As already noted above, the software used and developed for the satellite reception complex at FSC KSC SB RAS consists of three parts: 1) reception, 2) preliminary and 3) thematic processing. Let's consider its structure and characteristics:

- The data receiving software was developed by SCANEX Company, a provider of a satellite receiving station. It provides full automation of all processes: downloading of orbital elements, receiving GPS time signals, scheduling based on given priorities for receiving data from various spacecrafts, managing multiple receiving servers, and so on. If necessary, the operator is provided with all the current information on the real-time process of the station, including the spacecraft flight schedule and the actual video stream from it (if possible, decoding it during the fly).

- The preprocessing software includes three standard packages for data of different satellite systems: TERRA/AQUA for the MODIS radiometer, Suomi NPP/NOAA-20 for the VIIRS, CrIS, ATMS radiometers, Fengyun-3A/B/C for the MERSI radiometer. All packages operate in the Linux/CentOS environment on a single server. Processing scripts are started after a specified time interval (15 minutes).
- Thematic data processing is performed in accordance with the tasks to be performed in parallel on two servers. In particular, for each received satellite image TERRA/AQUA, Suomi NPP/NOAA-20 a set of multi-scale multichannel overview images / "Quick-looks" (popular spectral channel combination, surface temperature, NDVI) generates for substantially instant display of information on the Web page. The cartographic web interface provides the user with a large set of visualization capabilities for remote sensing data.

The range of tasks to be solved on the basis of the data of RSS is constantly expanding. Investigations in the development of real-time satellite monitoring systems, carried out at FSC KSC SB RAS, involves the creation of new and, at the first stage, the use of existing available programs for RSS processing [25], including algorithms for radiometers MODIS and VIIRS: search of active fires, evaluation of aerosols content, evaluation of fume area, masking cloudiness, determination of the optical properties of cloudiness (temperature, height, fraction and aggregate state of the upper limit), atmospheric correction of visible and near IR channels, recovery of vertical atmosphere profiles, calculating NDVI and EVI.

Conclusion

The significant increase in the amount of satellite data, observed at the present time, leads to the need of developing methods and technologies for the systems of receiving, storing and processing satellite information. The analysis of the developed solutions and our R&D experience shows that one of the most promising approaches in creating systems of this type is the implementation of a set of interconnected autonomous functionally-limited components that interact with each other in a service-oriented architecture with a user interface built on the basis of geo-information web-system technologies.

References

1. Duccio Rocchini, Vaclav Petras, Anna Petrasova, Ned Horning, Martin Wegmann. Ecol. Informatics, **40**, 57-61 (2017)
2. Yan Ma, Haiping Wu, Lizhe Wang, Bormin Huang, Wei Jie. Future Gen. Comp. Syst., **51**, 47-60 (2015)
3. Weipeng Jing, Dongxue Tian. Procedia Comp. Sci., **129**, 238-247 (2018)
4. F. Casu, M. Manunta, P. S. Agram, R. E. Crippen. Rem. Sens. of Env., **202**, 1-2 (2017)
5. Roberto Giachetta. Comp. & Graph., **49**, 37-46 (2015)
6. Kashnitskii, A.V., Lupyan, E.A., Balashov, I.V., Konstantinova, A.M., Atmos. & Ocean. Optics, **30**, 1, 84-88 (2017)
7. Craglia, M., de Bie, K., Jackson, D., Pesaresi, M., et al., Int. Journal of Digital Earth, **5**, 4-21 (2012)
8. Federico Bunkheila, Marco Cuollo, Emiliano Ortore. Acta Astronautica, **91**, 157-165 (2013)

9. Matthias Müller, Lars Bernard, Daniel Kadner. ISPRS J. of Photogr. and Rem. Sens., **83**, 193-203 (2013)
10. Konstantinos Evangelidis, Konstantinos Ntouros, Stathis Makridis, Constantine Papatheodorou. Comp. & Geosci., **63**, 116-122 (2014)
11. Adam Lewis, Simon Oliver, Leo Lymburner, Ben Evans, Lan-Wei Wang. Rem. Sens. of Env., **202**, 276-292 (2017)
12. Mojtaba Karami, Kazem Rangzan, Azim Saberi, Comp. & Geosci., **60**, 23-33 (2013)
13. Yanbo Huang, Zhong-xin CHEN, Tao YU, Xiang-zhi HUANG, Xing-fa GU. J. of Integr. Agricult., **17**, 1915-1931 (2018)
14. Weijie Liu, Bai Zhang, Zongming Wang, Kaishan Song, Jia Du. Procedia Env. Sci., **2**, 906-913 (2010)
15. L. Bastin, G. Buchanan, A. Beresford, J. -F. Pekel, G. Dubois. Ecol. Inf., **14**, 9-16 (2013)
16. Chandi Witharana, Daniel L. Civco, Thomas H. Meyer. ISPRS J. of Photogram. & Rem. Sens., **87**, 1-18 (2014)
17. Indranil Misra, S. Manthira Moorthi, Debajyoti Dhar, R. Ramakrishnan. Procedia Comp. Sci., **46**, 812-819 (2015)
18. USGS EarthExplorer // https://earthexplorer.usgs.gov/
19. Scanex UniScan Receiving station // http://www.scanex.ru/station/uniskan/
20. Yakubailik O.E., A. Kadochnikov A.A., Tokarev A.V. *International Multidisciplinary Scientific GeoConference SGEM2015 Conference Proceedings*. **2**, 1, 487-494 (2015)
21. Yakubailik O.E. Journal of Siberian Fed. Univ. Eng. & Tech., **9**, 7, 979-986 (2016)
22. Yakubailik O.E. *International Multidisciplinary Scientific GeoConference SGEM2016 Conference Proceedings*. **1**, 657-664 (2016)
23. NetApp Hybrid-Flash Storage System FAS2554 // https://www.netapp.com/
24. Extreme Networks // https://www.extremenetworks.com/product/x620/
25. Coronado P., Brentzel K. *Earth Science Satellite Remote Sensing* **2** (Beijing: Tsinghua University Press, 2006)

Software for data visualization in the system of real-time satellite monitoring

Oleg Yakubailik[1,*], *Alexey* Kadochnikov[1], and *Alexey* Tokarev[1]

[1] Federal Research Center "Krasnoyarsk Science Center of the Siberian Branch of the Russian Academy of Sciences", 50/45 Akademgorodok, Krasnoyarsk 660036, Russia

Abstract. Methods for creating multi-layered mapping representation for the real-time satellite monitoring systems, based on web-GIS technologies, are considered. The generated map can contain a set of raster and vector thematic layers of spatial data, additional information layers of images from below (a basic map – a substrate) and from above (a layer of labels) of user's thematic data. The client software provides interactivity of the created maps in a standard web browser. The configuration of the data presentation provides the ability to select spatial data format according to the client (vector or raster), control various display options (interactive legend, list of fields displayed in attribute tables). It also provides application of custom templates to display table information on layer objects, arrangement of data access rights - from opportunities to visualize a layer or map through a web browser to providing direct access to them through standard mapping WMS/WFS web services, web-based editing, or shape-file downloads. An integral part of thematic web maps, based on the geo-portal, is a customized rasterization and data caching service that provides a noticeable performance increase for "heavy" multi-layer maps. A set of mapping substrates, such as schematic maps, satellite images mosaics, digital terrain models, spatial data of outside web services are a good addition to maps on geoportal.

1 Introduction

Development of land remote sensing systems is currently conditioned by a combination of such factors as an increase in the number of spacecrafts and improvement of their performance characteristics, an increase in the availability of satellite information related to the development of the Internet: increase in speed and decrease in cost of access, the development of Web services and standards for the transfer of geospatial data [1]. New technologies for automation of receiving and processing remote sensing data, creation of super-large archives in specialized data centers have made it possible to arrange fundamentally new ways for satellite data operations, and to create a new generation of systems for real-time monitoring of the natural environment [2-4]. They are based on the geo-information web technologies, interactive online mapping systems with direct access to satellite information, provided with the ability to configure various parameters for visualizing geospatial data and generating complex analytical queries.

Technologies of geo-information web-systems and services are dynamically developing now [2-4]. One can find more than a hundred Russian geo-portals, dozens of geo-services, web GIS, mapping services and data visualizers, including specialized solutions for satellite data operations. Along with the traditional tasks of ordinary visualization of terrain maps and satellite images mosaics, up-to-date systems provide the user with the flexibility to customize various parameters of mapping visualization, spatial data processing services [5-7].

The authors' research experience in the subject under consideration shows that a geo-information platform, which conceptually determines the architecture and capabilities of implemented solutions, acts as the core element in the development of software for monitoring the state of the environment on the basis of land remote sensing data. The basic trend now is the use of the workplace of the GIS solution' end user, based on a web browser, as well as embedding the necessary set of scripts to work with the system in geo-information Internet portals. In some cases, these solutions are auxiliary and perform mostly spatial data viewing functions [8-10].

2 Geoinformation platform

Up-to-date web GIS are built in the so-called service-oriented architecture, and they can be considered as a set of interrelated software tools for arranging spatial data such as import/export, cataloging, visualization, creation, processing, distribution, etc. The technological basis for such solutions is usually the libraries of software interfaces such as Google Maps API, Mapserver Mapscript, ScanEx GeoMixer, etc. [11, 12]. They provide access to the functions and context of the mapping elements of the web pages: map visualization tools, spatial metadata, for example a road network with traffic jam data, as well as terrain, points on the map, etc. [13, 14]. The software and hardware support of the satellite monitoring system is also developed in accordance with this approach, because it also represents a web-based GIS.

One of the first tasks in designing any application system is the choice of a technological arrangement (geo-information platform) for GIS development, including software, technologies for storing spatial data and processing them [15-17]. Currently, there are traditionally considered two alternatives in the choice of basic software: based on commercial or free (open source) software products. Without considering financial, philosophical and short-term aspects of this choice, today there is often practically no opposition between the two approaches considered. And the reason is that commercial and free GIS are now complementary thanks to the compatibility of data formats, information exchange standards based on Web services, etc. For example, it is possible to analyze spatial data in ESRI ArcGIS, then convert them to MapInfo for further delivery to the customer, using free Mapserver software for presentation on web pages, and form a spatial metadata catalog using GeoNetwork Open Source. At the same time, it is possible to use the open source PostgreSQL DBMS with the PostGIS extension module in order to store spatial data. For most of the tasks is almost equal in performance and functionality to ESRI ArcGIS Server or Oracle DBMS with the extension for working with spatial data Oracle Spatial. Thus, the competence of specialists who provide hardware support for the solutions becomes a determining factor here. Commercial products are usually simpler in the initial configuration, they immediately from the start. Free software also has its advantages, related to the openness of architecture and expansion capabilities.

Pointing to the specific features of satellite monitoring systems, most of the algorithms for processing satellite data, developed in NASA, are available just in the source code. Specialists from all over the world are showing considerable interest in such resources, they can directly analyse existing methods of data processing and create their own modified algorithms. NASA is interested in constructive criticism and prompt correction of noticed errors.

3 The system for satellite data visualization

A geo-information web server was used as a software base for the catalog of satellite data. Its modular architecture allows developing the system already in the process of its operation, by adding new means of data access and also improving existing ones. The development is performed on the basis of free and open source software both in the GIS part and in all other development components. The described web application provides the user with means for filling and editing system data and metadata, searching and classifying the cartographic resources, analytical processing capabilities [18].

For the direct access to spatial data the Open Geospatial Consortium (OGC) protocols, such as Web Map Service (WMS), Web Map Tiling Service (WMTS) and others are used; they allow access to the geoportal resources from applied user programs directly such as ArcGIS, MapInfo, QGIS etc., as the support of these protocols is integrated into most of the modern geoinformation systems.

The server part of the system is implemented using the MVC design pattern (model-view-controller) and "Web 2.0" technologies [19]. Using this architecture involves splitting application data, the user interface, and control logic into three separate components: a model, a view, and a controller, so that each component can be modified independently. In the conditions of constant modernization of the system, continuous specification of technical requirements and statement of the task, these possibilities become very actual.

During the development of the system under consideration, several new software components and libraries were created that can be used (replicated) in other projects [20-22]. These are elements of the user interface, services for working with geospatial DBMS, applied cartographic mapping Web services, etc.

The basic functionality of the web GIS is provided by the software tools UMN Mapserver and MapGuide Open Source. Geospatial DBMS PostgreSQL/PostGIS, web content management system based on Drupal CMS and DokuWiki wiki system, a family of libraries for development of user interface and other system components for JavaScript programming languages, PHP – JQuery, Fusion, TinyMCE, ExtJS, MapScript, and many others are also used.

4 Satellite data catalog implementation

The priority investigation task at this stage was the development of software tools for working with satellite data catalogs. The developed software tools provide the solution of the priority tasks for the real-time processing of remote sensing data from the new UniScan satellite receiving complex at the FSC of the KSC of the SB RAS, which was put into operation in the spring 2017.

The basic software of the satellite receiving complex at FSC KSC of the SB RAS forms a set of standard products of the 1st level – channel images (brightness of spectral channels). Further tasks related to the extraction of useful information, thematic processing of data, are solved separately by means of special software.

The system of web-visualization of satellite information is based on sets of specially created collections of multi-scale images with the possibility to select on the web interface combinations of displayed channels on small scales and at the same time the availability of detailed data at the maximum available spatial resolution. The limitation in the choice of channel combinations at a detailed level is concerned with the saving of disk space. Thus, a compromise is realized between the class system "we work with swift examination and the system of the level "anything with maximum detailing". The created web interface provides viewing of the archive of satellite images. It implements the ability to select a spacecraft sensor, a predefined set of channels and products from each photo image for simple data analysis. The capabilities of the developed software module allow combining any combination of channels available in the image without additional configuration of the server software.

The modular architecture of the system under consideration, the use of standard web services for data exchange between these modules, ensures the rapid adaptation of the existing software to the customer requirements, the replication of individual compo-nents, and their complementarity. In its turn, registration of created resources (satellite images), i.e. spatial data in the centralized catalog, provides the possibility of their simultaneous application in several developments. This service-oriented approach, based on the active implementation of web technologies into application information systems, is increasingly being used now.

LUT (Look Up Table) is used for creation of color images. This is a kind of "table of corrections" for making changes to each of the three channels. Previously, in order to improve image quality, a spectral transformation was used, which is based on working with a spectral diagram showing the relationship between the number of the image pixels and the spectral brightness values. The image contrast parameter changes during spectral transformations. To increase the contrast, a linear stretching of the histogram was used. It means that new values are assigned to all values of brightness in order to cover the entire possible range from 0 to 255. LUT allows you to change the brightness of image points when converting the image to 8 bits of 16 bits by linear interpolation between them. This enables the picture to become more bright. An example of a web interface is shown in Fig. 1.

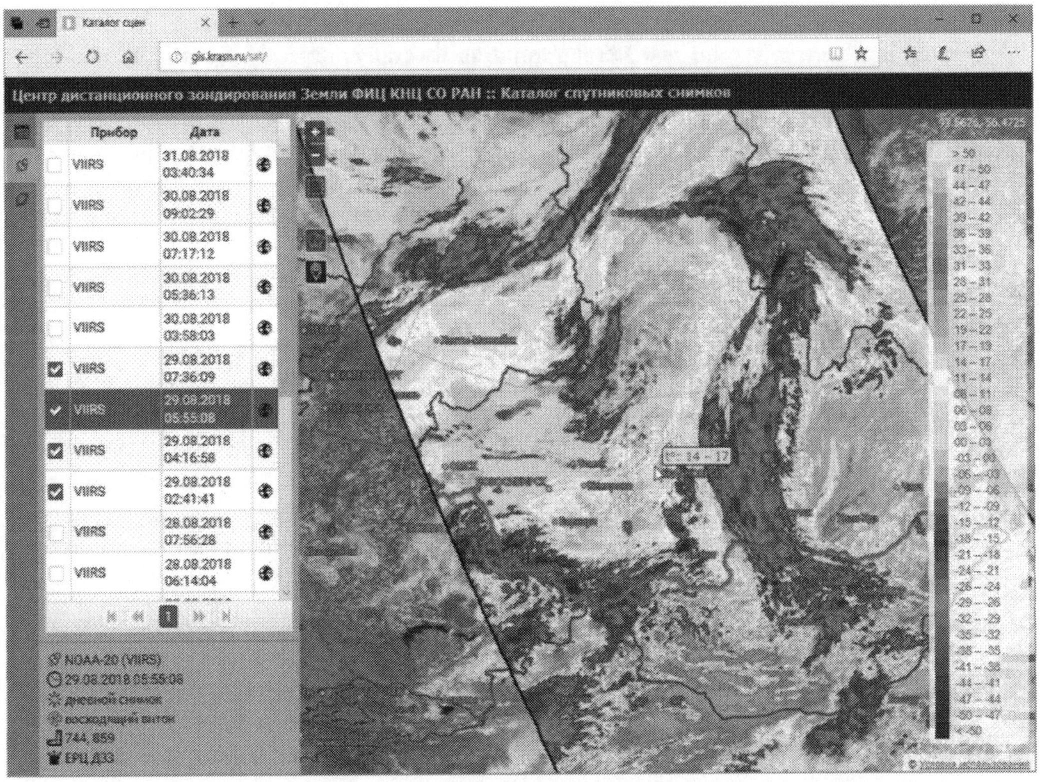

Fig. 1. The interface of the satellite images catalog of FSC KSC SB RAS

The basic tasks of development at this stage are the implementation of methods and algorithms for processing satellite data. The technological features of dealing with raster data are predetermined by the need to quickly display large files. The experiments with various formats and their parameters have shown that the optimal choice in terms of minimizing the display time in a web browser is the TIFF format with geolocation (GeoTIFF), using the TILES structure (the image consists of a set of independently stored one-type fragments), previously calculated pyramidal overview layers OVERVIEW. All the above processing steps are performed by the GDAL library utilities.

The developed software tools provide the solution of the priority tasks of the real-time processing of incoming remote sensing data, their cataloging, and interactive visualization using a web application. The following data were considered: MODIS TERRA/AQUA, SUOMI NPP/NOAA-20.

A set of server applications for the preliminary processing of satellite data was prepared for quickly and easily search in the catalog of satellite data and for minimizing the load on the server software and hardware. It includes the following processing steps:

1. Conversion of the original raster data to GeoTIFF format with the transformation of the original projection into the Lambert Azimuthal Equal Area (EPSG code: 3576), which is mandatory for the Open Geospatial Consortium services.
2. Creation for all scenes a catalog of color images in coarse resolution ("swift examination") in raster PNG format with transparency. Such images are used for quick overview of satellite data registered in the catalog with simultaneously visualization of a set of such ("swift examination") in a web browser window, with scaling on the client side, without reloading/retrieving data via the Internet.
3. Creation of multi-scale set of multi-channel images of GeoTIFF format. Each image, created in a certain spatial resolution, contains several practically significant channels, from which a certain set of color composite images are generated in the web application in parallel with the background processes. For example, for the TERRA/MODIS sensor, the multi-channel image contains 1 , 2, 3, 4, 7, 31 channels, and on this basis a set of commonly used composite RGB images is dynamically formed: 1-4-3 ("natural colors"), 7-2-1 ("Fires"), 3-6-7 ("Snow and Ice"), as well as single-channel images with palettes – 31 ("Temperature"), 2-1/2+1 (NDVI) .

Conclusion

Development of the effective software and hardware tools for the tasks of regional satellite monitoring based on the technologies of geo-information web systems has the significant importance. The developed approach can serve as a basis in solving actual applied problems related to the current and future needs of the Krasnoyarsk Region, based on use of land remote sensing data, up-to-date infrastructure of spatial data for effective social and economic progress and innovation development, improving competitiveness and ensuring safe living.

References

1. Kashnitskii, A.V., Lupyan, E.A., Balashov, I.V., Konstantinova, A.M., Atmos. & Ocean. Optics **30**, 1, 84-88 (2017)
2. P. Soille, A. Burger, D. De Marchi, et al. Fut. Gen. Comp. Syst., **81**, 30-40 (2018)
3. Körting T.S., Fonseca L., Câmara G. Comp. & Geosci., **57**, 133-145 (2013)
4. Kumar S., Peters-Lidard C., Tian Y., et al. Env. Model. & Soft., **21**, 1402-1415 (2006)
5. Pinde Fu, Jiulin Sun. *Web GIS: principles and applications* (Esri Press, 2011)

6. Granell C., Díaz L., Gould M. Env. Model. & Soft., **25**, 182-198 (2010)
7. Gorelick N., Hancher M., Dixon M., Ilyushchenko S., Moore R. Rem. Sens. of Env., **202**, 18-27 (2017)
8. Singh M., Gautam R., Gatebe C., Poudyal R. Comp. & Geosci., **96**, 173-180 (2016)
9. Doneus M., Verhoeven G., Atzberger C., et al. J. of Archaeol. Sci., **52**, 84-96 (2014)
10. Kirschbaum D., Stanley T., Zhou Y. Geomorphology, **249**, 4-15 (2015)
11. Li Songnian, Suzana Dragicevic, Bert Veenendaal. *Advances in Web-based GIS, Mapping Services and Applications* (CRC Press, 2011)
12. J.T. Sample, K. Shaw, S. Tu, et al. *Geospatial Services and Applications for the Internet* (Springer-Verlag US, 2008)
13. Mari R., Bottai L, Busillo C, Calastrini F. et al. Renewable Energy, **36**, 754-763 (2011)
14. Rinner C., Keßler C., Andrulis S. Comp., Environ. and Urban Syst., **32**, 386-395 (2008)
15. Paciello R., Coviello I., Bitonto P., et al. Env. Model. & Soft., **77**, 50-62 (2016)
16. Agapiou A., Lysandrou V., Alexakis D., et al. Comp., Env. and Urban Syst., **54**, 230-239 (2015)
17. Hannemann W., Brock T., Busch W. Int. J. of Coal Geol., **86**, 54-57 (2011)
18. David B. Lobell. Rem. Sens. of Env. **164**, 324-333 (2015)
19. L. Bastin, G. Buchanan, A. Beresford, et al. // Ecol. Inform., **14**, 9-16 (2013)
20. Yakubailik O.E., Kadochnikov A.A., Tokarev A.V. Optoelectr., Instr. and Data Proc., **54**, 243-249 (2018)
21. Yakubailik O.E., A. Kadochnikov A.A., Tokarev A.V. *International Multidisciplinary Scientific GeoConference SGEM2015 Conference Proceedings*. **2**, 1, 487-494 (2015)
22. Shaparev N., Yakubailik O. MATEC Web of Conf., **79**, 01081 (2016)

The Intelligent Managerial Decision Support System for Agricultural Land Evaluation

Yury Maglinets[1], *Ksenia* Raevich[1,*], and *Gennady* Tsibulsky[1]

[1] Siberian Federal University, Institute of Space and Information Technologies, Chair of Systems of Artificial Intelligence, 660074 Kirenskogo St. 26, Krasnoyarsk, Russia.

Abstract. The article presents the structure of the intelligent managerial decision support system for agricultural land evaluation. The assessment problem has been a technique of defining performance indicators of productive agricultural land with account for geographical and regional features of its configuration and infestation that should be considered in making decisions during field operations. The indicator evaluation scheme is given. The specific example of evaluation is stated. Land cultivating deficiency depending on its overgrowing by trees and shrubs is shown. The results obtained are proved to be used in managing an agricultural sector of the region.

1. Introduction

Analyzing land as a component of the country's national wealth is a comparable quantitative and economic evaluation of its usability and the ecological and economic effects of using land plots as an object of evaluation for different purposes.

An objective basis for solving evaluation problems is knowledge about the object of evaluation as a spatially localized part of the earth's surface, characterized by a set of measurable factors. On the one hand, land, as a natural resource, is characterized by location in space, relief, soils, plant and animal life. It is evaluated as its possibility of performing multi-purpose functions [1-4]. On the other hand, land, as an object of economic relations, is evaluated in terms of its utility and profitability from using a particular plot [3]. Besides, agricultural land evaluation is influenced by economic, physical and social factors, location aspects (distance to communities, water bodies, etc.), transport network development, etc. Under the influence of these factors, the demand for land plots increases or decreases and their market prices are adjusted.

This way, agricultural land, as an object of economic relations, has use and market value, which is defined by its agricultural land evaluation, as well as the land occupied by buildings, structures used for the production, storage and primary processing of agricultural products [4].

[*] Corresponding author: ksenia_248@mail.ru

The system considered in this paper allows us to formulate assessments of the state of agricultural land, as well as carry out information support for decision-making in managing land resources.

2. The functional structure of the intelligent managerial decision support system

The intelligent managerial decision support system is composed of the following subsystems:
1) Agricultural land evaluation has a complex, multifactor analysis. In general, it is necessary to take into account the influence of climatic factors, characteristics of soil and vegetation cover, infrastructure features, geospatial characteristics of the estimated areas of the Earth's surface, etc. *The knowledge base* is the key block of the system.
2) *The knowledge base editor* allows the expert responsible for the ontology content to perform interactive operations on creating a new ontology (ontology version) and editing it.
3) *The problem solution generator* allows the expert responsible for the problem to create and maintain a problem description model within a certain version of the ontology, and also perform operations to migrate the problem description into updated versions of the ontology.
4) When creating and modifying an ontology, there is a procedure describing computational relationships for evaluation metrics. This procedure is based on using *the formula editor* that is a component of the knowledge base editor and the problem solution generator. The formula editor contains basic math operations and is able to build table functions.
5) *The geospatial data editor* serves to prepare and operate a geospatial description of the instance of the problem being solved.
6) While creating an empty instance of the problem being solved, *the upload module of the problem solution generator* generates a vector layer template with a predefined set of attributes, specified in the problem solution scheme. The template is placed in the geospatial database.
7) *The geospatial database* is a set of layers filled with data on the spatial coordinates of evaluation objects.
8) Attribute data contains the values of the actual parameters of evaluation objects. *The data import-export module* is responsible for data transfer. This module interacts with an agricultural monitoring automation system [6] and other external systems, which allows receiving data based on the processing and analysis of satellite imagery, ground measurements, weather station data and other sources.
9) *The evaluation module* does interim and final evaluation.
10) Evaluation visualization is carried out through *the calculator interface.*

3. Approbation

The system has been tested in solving performance evaluation problems of using agricultural land with analyzing the degree of overgrowing by trees and shrubs.

It is quite natural that to improve the agroeconomic performance of agro-industrial enterprises it will be expedient to increase the agricultural land area by carrying out a complex of reclamation work on overgrown cropland [3].

At the first stage of the methodology, the problem of evaluating the technological efficiency coefficient of the agricultural contour (TECAC).

The AC ontology is given by a quadruple:

$$O =< K, T, E, M > \quad (1)$$

where K is the taxonomy of the evaluation criteria, T is the set of solved problems, E is the set of metrics for characteristics evaluation, M is the set of primary metrics that allow evaluating the numerical value of the characteristic in physical term.

The heart of the ontological model under consideration is the K taxonomy of the TECAC:

$$K =< N, R > \quad (2)$$

where $N = \{ni\}$ is the set of taxonomy classes – attributes of the evaluation object, $R \subset N \times N$ is N order relation.

At the second stage of the methodology, the formula module forms the computational procedure for the TECAC.

The TECAC basic procedure consists in calculating the ratio of the current technological efficiency of the AC to the maximum possible.

$$K_{TECAC} = TECAC_C / TECAC_T \quad (3)$$

where K_{TECAC} is the technological efficiency coefficient of grain production for the AC;

$TECAC_T$ is the total technological efficiency of the AC, evaluated without trees and shrubs;

$TECAC_C$ is the current technological efficiency of the AC, evaluated with trees and shrubs as pegs.

In general, the TECAC is evaluated by the following ratio:

$$TECAC = PI - PC \quad (4)$$

where PI is the projected income and PC is the projected cost of grain production, which includes the cost of fertilization, seeds, fuel and lubricants, salaries and wages, annual depreciation and service of grain harvesting machines, etc.

When going around a plot of trees and shrubs, the path length is calculated as the arc length of the cut-off sector s. For each subsequent roundabout, the height of the cut-off sector h increases, thereby increasing the length of the arc s.

Table 1. Cost indicator values in grain production

Index	Estimated value
Fuel consumption per hour	$C_{hour} = C_{sp} * P_{en} = 162 * 235 = 38.07$кg / h C_{sp} – specific fuel consumption, gr / (horsepower-hour) P_{en} – engine horsepower
Fuel consumption per 1 ton of threshed grain	$C_c = C_h / HC_h = 38.07/14$=2,72 kg / t HC_h – harvester capacity per hour, t / h
Consumption of fuel and lubricants per 1 ha, at a given yield	$C_{ha} = C_f * GY_{ha} = 2.72 * 2.3 = 6.26$ кg / ha C_f – fuel consumption per ton, kg / t GY_{ha} – grain yield per hectare, t / ha For the region considered, the mean value is 23 dt / ha.
Expenses for fuel and lubricants in value terms	$C_T = C_f * C_{df} = 2.72 * 35 = 95.2$ RUB / t C_f – fuel consumption per ton, kg / t C_{df} – cost of diesel fuel, rubles / kg Calculated as the producer price of 1t of diesel fuel, which was 35.000 rubles per ton in 2015. $C_{ha} = C_{ha} * C_{df} = 6.26 * 35 = 219.1$ RUB/ha C_{ha} – fuel consumption per 1 ha, kg / ha C_{df} – cost of diesel fuel, rubles / kg

At the third stage of the methodology, the cost indicator values in grain production (Subproblem 1) are evaluated (see Fig. 1).

At the fourth stage of the methodology, the evaluation module calculates the total and current technological efficiency of the AC with the use of cost indicators from Table 1:

$$\mathrm{TECAC_t} = \mathrm{PI} - \mathrm{PC} = 23000000 - 219100 - 667000 - 332810$$
$$-3408000 = 18373090\ \mathrm{RUB}$$
$$\mathrm{TECAC_c} = \mathrm{PI} - \mathrm{PC} = 20700000 - 197190 - 8389.5 - 600300 - 299529$$
$$-3062700 = 165131891.5\ \mathrm{RUB}$$

The TECAC coefficient is equal to:

$$\mathrm{K_{TECAC}} = \mathrm{TECAC_c}\ /\ \mathrm{TECAC_t} = 0.89 * 100\% = 89.95\%$$

This means that the AC (Subproblem 1) uses 89.95% of its agro-economic capability.

At the fifth stage of the methodology, the evaluation module calculates the TECAC coefficient for different degrees of overgrowing by trees and shrubs in Table 2.

Table 2. The TECAC coefficient for AC with trees and shrubs

Index	Estimated value for AC with trees and shrubs,%					
	0	10	20	40	60	80
The cultivated area of AC, ha	1000	900	800	600	400	200
Projected income from grain marketing, million rubles	23	20.7	18.4	13.8	9.2	4.6
Costs for fuel for harvesting (without idling while driving around the pegs), rubles	219100	197190	175280	131460	87640	43820
Costs for fuel at idling while driving around the pegs, rubles	0	8389.5	16779	33558	50337	67116
Depreciation and service, thousand rubles	667	600.3	533.6	400.2	266.8	133.4
Salaries and wages, rubles	332810	299529	266248	199686	133124	66562
Other costs, thousand rubles	3408	3062.7	2726.4	2044.8	1363.2	681.6
Losses, million rubles	-	1.85	3.69	7.38	11.08	14.77
$\mathrm{K_{TECAC}}$, %	-	89.95	79.91	59.81	39.72	19.6

The data, calculated by the TECAC, show the volume of financial losses for different degrees of the AC overgrowth by trees and shrubs.

The scale of acceptable and unacceptable losses is determined by the expert in Table 3, based on the agro-economic expediency of the decisions made.

Table 3. The scale of loss acceptability of the AC overgrowth by trees and shrubs

Index	Acceptable losses	Unacceptable losses
Losses, million rubles.	≤3.69	>3.69
$\mathrm{K_{TECAC}}$, %	≥79.91	<79.91

At the sixth to ninth stages, the results obtained are compared to make the necessary managerial decisions. The calculator interface provides the decision maker with the results.

Conclusion

The paper considers the intelligent decision support system that allows solving the problems of ranking agricultural land based on their characteristics and agronomical GIS data. The technique of determining technical and economic indicators for cultivating agricultural fields in the system for managing the efficiency of the agro-industrial complex of the region is formulated. As a basis for determining technical and economic indicators, an example of calculating the technological efficiency coefficient of an agricultural contour using an intelligent system for agricultural land evaluation is given [4]. The technological efficiency coefficient calculates the impact of negative processes, expressed in the AC overgrowth by trees and shrubs, on the AC capacity.

At present, the methodology is being tested in solving the problem of complex agro economic land evaluation by the example of the Sukhobuzimsky district of the Krasnoyarsk Territory. The technological efficiency coefficient of the AC is one of the factors of the AC evaluation [4].

The reported study was funded by Russian Foundation for Basic Research (No.18-47-242002), Government of Krasnoyarsk Territory, Krasnoyarsk Regional Fund of Science, to the research project: «The development of technology for creating intelligent information systems of object-oriented monitoring of territories based on remote sensing data».

References

1. K.V. Shatrova, Y.A. Maglinets. The decision support system for agricultural land evaluation // The current issues of Earth remote sensing from space: collection of abstracts of the 12th Open Russian Conference (Moscow, November 10-14, 2014). M.: Space Research Institute, 2014. P. 394.
2. K.V. Shatrova, Y.A. Maglinets, G.M. Tsibulsky. The view model for agricultural land state and dynamics data // Journal of Siberian Federal University. Series: Engineering and Technology. 2014. V. 7. No. 8. P. 984–989.
3. I.V. Zenkov, E.V. Loginova. Managing the innovative development of the land sector of the agro-industrial complex in the Krasnoyarsk Territory: monograph. Krasnoyarsk: SibFU, 2013. 236 p.
4. K.V. Raevich, I.V. Zenkov, Y.A. Maglinets. Managing the productive agricultural land of the Krasnoyarsk Territory agro-industrial complex using the agro economic potential factors // Bulletin of Irkutsk State Technical University. 2016. No. 3. P. 57–66.
5. The functional subsystem “The electronic atlas of agricultural land”. EPassport FS-7713 No 0296 from 07.02.2013. URL: http://atlas.mcx.ru (29.03.2016).
6. K.V. Shatrova. Spatial data representation for the best route search area organization / K.V. Shatrova, Y.A. Maglinets // International Scientific Conference “The regional issues of Earth remote sensing” / science editor E.A. Vaganov; publications assistant. A.V. Mashukova – Krasnoyarsk : SibFU, 2014. – 366 p. P. 345 – 348.

AUTHOR INDEX